THE FUTURE OF WINTER CITIES

Volume 31, URBAN AFFAIRS ANNUAL REVIEWS

THE FUTURE OF WINTER CITIES

Edited by
GARY GAPPERT

Volume 31, URBAN AFFAIRS ANNUAL REVIEWS

SAGE PUBLICATIONS
The Publishers of Professional Social Science
Newbury Park Beverly Hills London New Delhi

This book is dedicated to Jane Jacobs and Jean Gottmann for their excellent writings that have deepened our understanding about the evolution and resurgence of cities in an increasingly complex global economy. Partial support for The Future of Winter Cities *should be credited to the Academic Affairs Office of the Canadian Embassy in Washington, D.C., and to the Northern Ohio Inter-Institutional Urban Demonstration Program.*

For information address:

SAGE Publications, Inc.
2111 West Hillcrest Drive
Newbury Park, California 91320

SAGE Publications Inc.
275 South Beverly Drive
Beverly Hills
California 90212

SAGE Publications Ltd.
28 Banner Street
London EC1Y 8QE
England

SAGE PUBLICATIONS India Pvt. Ltd.
M-32 Market
Greater Kailash I
New Delhi 110 048 India

Printed in the United States of America

Library of Congress Cataloging in Publication Data

The Future of winter cities

(Urban affairs annual reviews; v. 31)

Includes bibliographies.

1. City planning—Environmental aspects—Addresses, essays, lectures. 2. Urban renewal—Addresses, essays, lectures. I. *Gappert*, Gary. II. Series. III. Title: Winter cities.

HT107.U7 vol. 31 [HT166] 307.7'6 86-1850
ISBN 0-8039-2620-1 [307.7'64]
ISBN 0-8039-2621-9 (pbk.)

FIRST PRINTING

CONTENTS

Introduction: The Future of Winter Cities 7

1. The Cold City: The Winter of Discontent? 13
Seymour Sacks, George Palumbo, and Robert Ross

PART I: THE CONDITION OF WINTER CITIES **35**

2. The Survival of Winter Cities: Problems and Prospects 49
Norman Pressman

3. Toronto: Policies and Strategies for the Livable Winter City 71
Xenia Zepic

4. Cities and Snow: Reflections from Finland 94
Jorma Mänty

5. Sapporo: A Japanese Winter City 100
Yasuo Masai

6. Anchorage: Toward an Explanation of Migration to Winter Cities 117
Knowlton W. Johnson, C. Theodore Koebel, and Michael L. Price

7. Minneapolis: The City That Works 131
Donald M. Fraser and Janet M. Hively

PART II: THE TRANSACTIONAL CITY: ISSUES AND CASES **139**

8. The Status of the Transactional Metropolitan Paradigm 145
Kenneth E. Corey

9. The Great Lakes Megalopolis 163
Harold M. Mayer

10. The Nation's Most Livable City: Pittsburgh's Transformation 173
Brian J.L. Berry, Susan W. Sanderson, Shelby Stewman, and Joel Tarr

11. Knowledge and the Advanced Industrial Metropolis 196
Richard V. Knight

12. The Arts and Urban Development 209
William S. Hendon and Douglas V. Shaw

13. Food Policies for Cities in Northern Latitudes 218
Jon Van Til and Medard Gabel

PART III: THE REVIVAL OF INDUSTRIAL CITIES **229**

14. Alternative Paths to the Revival of Industrial Cities 233
Wilbur R. Thompson and Philip R. Thompson

15. Urban Development in an Advanced Economy 251
Royce Hanson

16. Integrating Urban and Industrial Employment Policies 267
John P. Blair and Tran Huu Dung

17. The Politics of Revitalization: Public Subsidies and Private Interests 285
Terry F. Buss and F. Stevens Redburn

18. The Revitalization of New England Cities 297
James M. Howell

19. Transforming a Manufacturing City: Akron's Redevelopment Plans 315
Frank J. Costa, Jack L. Dustin, and James L. Shanahan

About the Authors 333

Introduction: The Future of Winter Cities

"My country is not a country; it's the winter."

—*Gilles Vigneault*

"Light must come from the north."

—*winter climate conference, Sapporo*

Winter has both a figurative and a literal meaning. The mention of winter calls forth a host of conflicting emotions and contradictory viewpoints. As one writer puts it, "poets traditionally equate winter with the passing of life, and find within the season metaphors for death." But even with this metaphorical definition, winter is always followed by spring.

Others might claim that the literal definition of winter is enough to evoke dread, if not fear. To them winter is a time to hibernate or to migrate, following waterfowl to some semitropical refuge. For them the bracing weather of winter is enervating rather than invigorating.

In this book about the future of winter cities, both definitions are applicable. Indeed, this book was originally conceived as a volume about the revival of industrial cities and the comeback of Snow Belt cities. But with the formation in 1984 of a Livable Winter City Association in Canada, the development of this urban affairs annual shifted to a more positive elaboration of the concept of "winter cities." According to the Livable Winter City Association (LWCA), at least 30 countries are located in the northern latitudes of the world and over 600 million people share the experience of living in winter climates. To the LWCA the term *winter city* encompasses most medium or small urban settlements above the 45-degree northern latitude.[1] (Unfortunately, this particular boundary would exclude such notable winter cities in the United States as Buffalo, Milwaukee, and all of New England. The 40-degree latitude that runs through Salt Lake City, Denver, and Philadelphia is perhaps more appropriate as a Snow Belt boundary if such a boundary is necessary.)

The LWCA seeks positive ways to revitalize northern cities, plus the means to capitalize on their affirmative values, and techniques to minimize the allure of Sun Belt areas. The LWCA feels that one cannot ignore the reality of a cold climate, but that one can generate ideas and programs either to mitigate winter or to take advantage of it. An alternative strategy is to emphasize the summer swelter of the Sun Belt as a counterbalance to the winter chill of the North. The group seeks not just to cope with, but even to "exalt" the essential nature of their climate. They want to design and rebuild for comfort and efficiency in cold climates, to develop and redevelop cities and towns to function optimally and to stress affirmative adaptations to real conditions. This would cover everything from urban form, architecture, and new products to lifestyle adaptation and enhancement of winter recreational opportunities.

In a broader sense, however, the revival and future of older cities in winter climates or northern latitudes should be viewed as a function of at least three forces.

First, the older industrial cities are now competing in a growing global economy and the process of deindustrialization—the migration of unskilled and semiskilled jobs to low-wage countries and regions—must be realistically understood as part of the development of an advanced industrial society.

Second, some of the older cities have evolved or matured very significantly as transactional cities and have developed new roles that are significant to the economy of the so-called information society and what is called high tech/high touch.

Third, some northern cities that are dominated by winter have aggressively demonstrated that new design initiatives can create a highly livable urban environment with world-class status such as that exhibited by skywalks in Minneapolis and Calgary and the underground malls in Montreal and Toronto.

WHITHER THE SUN BELT?

John Naisbitt, in his notable book *Megatrends*, has told us flatly that the population migration in the United States from the Snow

Belt to the Sun Belt is "irreversible in our lifetime." Perhaps so. But all population migrations include a back or reverse migration, usually of a lesser magnitude. It may well be that the explosive growth of Sun Belt cities in the last decade, largely fed by the highly mobile baby boom population that reached young adulthood in the 1970s and 1980s, will damper significantly as we approach the twenty-first century. There is also some indication that as the members of that baby boom population settle down and begin to have children, there is some return to areas in which they have relatives, grandparents, friends, and support networks.

In *Cities in the 21st Century* (Gappert and Knight, 1982), Butler and Chinitz point out that by the year 2000, the eight western states east of the West Coast will increase their population by 50% but that will be only 5 million people out of a projected U.S. population increase of over 40 million. They indicate that by 2030, "competition for scarce resources—water and land, in particular—will become a predominant factor in determining the economics and characteristics of development." There are limits to growth. There is an old German saying that "trees don't grow to the sky." No trend goes on forever and it is unlikely that the quality of life in so-called Sun Belt regions will persist undiminished into the next century.

Migration alters the nature of both the place of outmigration and the place of immigration. Denver's brown cloud, the Houston school system, and the homicide rates in Atlanta and Miami are all becoming impediments to the future growth of Sun Belt cities.

THE DEVELOPMENT OF SOUTHERN CITIES

In their path-breaking urban affairs annual, *The Rise of the Sunbelt Cities*, Perry and Watkins (1977) have elaborated the socioeconomic development of cities in the American South and Southwest. They describe a process of regional change characterized by a pattern of uneven urban development. In the postwar phase of capital accumulation, the six new pillars of growth—agriculture, defense,advanced technology, oil and natural gas, real estate and construction, tourism, and leisure—have all emerged with a dominant presence in Sun Belt locations. Various forms of federal assistance and support to these growth industries were uniquely beneficial to the stimulation of growth in southern and southwestern states, primarily in cities. They concluded that it was the impact of the various federal policies that was "the single most significant factor contributing to the sudden ascent of these cities" (Perry and Watkins, 1977: 50).

Elsewhere in the same volume Perry and Watkins indicate that in their view, the rise of the Sun Belt cities represents a geographical shift in the successful practice of the city as the dominant unit of profit. They foresee a time when the "advantageous" low-income pool of southern workers will be transformed into a poverty-stricken underclass, which will enable them to achieve a "crisis parity" with northeastern cities. They conclude that the rise of the Sun Belt cities does not represent the development of any new urban function in the United States but just a regional reaffirmation of the prevailing urban function, which is to utilize low-income and low-skill workers efficiently and effectively (Perry and Watkins, 1977: 292–293).

More recently Jane Jacobs in her *Cities and the Wealth of Nations* (1984) has outlined and elaborated a theory about the resurgent vitality of urban regions that possess an export replacing industrial ingenuity. She tries to demonstrate how economic life expands and develops as cities replace their imports with goods and services of their own production and then improvise innovations on earlier foundations. In her view, cities are forever losing older work and their vitality depends upon their ingenuity at developing new products, new services, and new markets.

Her model of vigorous import-replacing cities is complementary to the concepts of Jean Gottman about the transactional city. Certainly, the vitality of both Boston and Baltimore are a reflection of how the apparent intangibility of what is called the amenity infrastructure can contribute to the realities of an urban resurgence.

That resurgence is not necessarily associated with any dramatic increase in population. A city does not have to be huge or crowded to be great. This is especially true for industrial cities experiencing their inordinate growth in the early decades of the twentieth century. It is said, for instance, that from 1910 to 1920 Akron was the fastest-growing city in the United States. But currently Akron no longer produces tires, its most famous product. Instead, as is discussed in one of the chapters in this book, Akron has become the home of a number of multinational *Fortune* 500 corporations, a large university, and the internationally acclaimed Ohio Ballet.

Writing in the recent publication of the Brookings Institution, *The New Urban Reality* (1985), Peterson writes:

> Industrial cities must simply accept a less exalted place in American political and social life than they once enjoyed. . . . Policies must adapt to this new urban reality.

Simultaneously, however, the new edition of *Places Rated Almanac* (Boyer and Savageau, 1985) ranks Pittsburgh number 1 in the quality of life of over 300 cities in the United States with Boston as number 2. The rebirth of these two winter cities, both notable manufacturing centers in the 1950s, goes back several decades when their population stopped growing dramatically and their civic leadership became more concerned with enhancing the quality of their institutional and cultural development.

When Roger Starr of the *New York Times* first advanced his concept of "planned shrinkage" in the mid-1970s, it experienced considerable denial and rejection. Today small may not yet be beautiful but rapid growth is seen more as a concern than as a desirable achievement, except perhaps by developers and other real estate interests.

The new concern for optimal size—if matched by a concern for working-class mobility (as reflected in the new federal five-year, billion dollar program of housing vouchers)—represents one aspect of the urban future that the civic culture of the older industrial cities must address as the 1990s loom ahead.

WHAT THEN OF WINTER?

Even as some industrial cities shrink and others evolve as cities with a preeminent transactional economy, all northern cities must still address the problems of climate. Here the examples of Toronto, Montreal, and Minneapolis are illustrative. The development of multibuilding, multiblock pedestrian walkways, either as second-story skywalks or as downtown underground malls, are a partial solution to the problem of winter. More significant new production processes for the mass production of glass at a low cost provide a new urban design challenge and opportunity for more climate-responsive buildings.

Peter Brobert, a Swedish architect, has called this development as important as the development of reinforced concrete was in its time, and exalts the possibility of "the continuous and connected city where buildings are joined together by an enclosed urban room and where optimal climatic zoning creates a whole new energy situation" (Brobert, 1985). He cites the potential contribution of these new "urban rooms" to the expansion of urban culture and to the future evolution of humanity as Homo Urbanus.

Before we review the future and present conditions of winter cities in Part II, this section will continue with an analysis of "cold cities" by that great urban empiricist Seymour Sacks and two of his colleagues from Syracuse University. Their analysis identifies the particular role of annexation in the growth of southern cities and indicates that the loss of employment in northern cities may have peaked in the last decade.

After reviewing the conditions and potentials of winter cities in Part II, the book turns to an analysis of issues and cases associated with the concept of transactional cities and transactional urbanization in Part III. In Part IV the contributors examine in more detail some of the developmental realities of industrial cities, both specifically and in general.

As we consider the future of winter cities, one is struck by the complexity of the challenges that confront urban planners, urban scholars, and urban officials as we approach the twenty-first century. It is hoped that this volume will contribute to the development of more innovative and insightful solutions to problems of cities in northern climates that will ensure that they have a warm future.

NOTE

1. A newsletter and other material from the Livable Winter City Association is available from Jack Royle, P.O. Box 5130, Station A, Toronto, Ontario, Canada, M5W. The concept of winter as a public policy issue is credited to William C. Rogers when he was serving as a member of the Minneapolis committee on Urban Environment in 1977.

REFERENCES

BOYER, R. and D. SAVAGEAU (1985) Places Rated Almanac. New York: Rand McNally.

BROBERG, R. (1985) "Foreward: all-year cities and human development," in N. Pressman (ed.) Reshaping Winter Cities: Concepts, Strategies and Trends. Ontario: University of Waterloo Press.

GAPPERT, G. and R. V. KNIGHT (1982) Cities in the 21st Century. Beverly Hills, CA: Sage.

JACOBS, J. (1984) Cities and the Wealth of Nations. New York: Random House.

NAISBITT, J. (1983) Megatrends. New York: Warner Books.

PERRY, D. and A. WATKINS (1977) The Rise of the Sunbelt Cities. Beverly Hills, CA: Sage.

PETERSON, G. (1985) The New Urban Reality. Washington, DC: Brookings Institution.

ROGERS, W. C. and J. K. HANSON (1980) The Winter City Book: A Survival Guide for the Frost Belt. Edina, MN: Dorn Communications.

1

The Cold City: The Winter of Discontent?

SEYMOUR SACKS, GEORGE PALUMBO, and ROBERT ROSS

□FOLLOWING THE FUROR over the rise of the Sun Belt cities, it is now time to assess the present circumstances of those older American cities that might well be described as winter cities. Although the cities of the Northeast and Midwest have in the past been associated with this group of cities, local circumstances may be at least as important as regional location. The success of the Sun Belt cities has in large part been related to overall metropolitan area growth and the ability of these central cities to capture population and employment from their surrounding areas through annexation. In contrast to this, the decline of America's winter cities is linked to their inability, due to state laws or the presence of incorporated governments in surrounding areas, to change their boundaries. When these cities with static boundaries are viewed in the context of moderately growing, or even declining, metropolitan areas, predictions of rebirth and resurgence become suspect.

This study outlines the changes in populations, households, employment, unemployment, and income that have occurred since 1960, for central cities and their outside areas using information from the various decennial and quinquennial censuses. It is from this perspective that the long-term decline of the winter cities and the changes that might be viewed as a source of revival are analyzed. The identification of winter cities and the analysis of their present circumstances includes an account of the intrametropolitan and interregional changes of population and economic activity.

CITIES WITH FIXED BOUNDARIES

The common characteristic in our classification of winter cities is not the relatively cold climate that many of them share, nor is it any other strictly geographic characteristic. Rather, we have decided to view the winter city as one that remains in relatively fixed boundaries and as such is out in the cold when it faces the forces of decentralization. Although this taxonomic device is seemingly free of any *ex ante* regional classification, those cities that have maintained stable boundaries are, with a few exceptions, generally located in the Northeast, upper Midwest, and scattered city-counties. These cities are old cities in the sense that many of them have been undergoing serious long-term declines in population and employment, and they might be seen as being in a winter stage of a life cycle. The nature of the decline, and the emergence of a revival among some of this group, have interesting implications for students and policymakers alike.

The basis for the classification of winter cities and non-winter cities can be seen in Table 1.1, which presents information for changes in central city (CC) and outside central city (OCC) population in 85 of the largest metropolitan areas in the United States. The SMSAs are held in constant 1970 definitions and the change in city land areas reflect annexation activity. Generally those central cities that have been able to annex land from their surrounding areas have been the most successful at increasing or at least maintaining their populations, and with a few exceptions those cities that have not had significant annexations have declined in population. In Table 1.1 the column headed "Old" has a one placed next to any city that had a lower population in 1980 than in 1960 and can then be considered old in the sense that they are past their prime. A zero in the "Old" column indicates that the 1980 population is larger than the 1930 population, although several of these cities have been declining since reaching more recent peaks.

A one in the column labeled "Fixed Boundaries" indicates that the central cities of the SMSA have not annexed 5% of their territory in the periods 1960–1972, 1972–1977, or 1977–1982. These are the cities with static boundaries. A zero in this column indicates annexing activity in the period between 1960 and 1982. Perhaps the most striking feature of this table is the large number of annexing cities distributed throughout every region but the Northeast. Of the 85 central city areas, 50 had substantial annexations over the time period. Of further interest is the close, but not unequivocal, relationship between our old cities and nonannexing places. The ability to annex surrounding territory is not insurance against population decline, as is evident from the population declines of annexing cities such as Dayton, Toledo, and Birmingham. In contrast, the strong growth of Miami, a city with unchanging boundaries, suggests there are sources of growth other than the suburbanization of the central city.

Although the exceptions are of some interest, the overwhelming evidence supports the idea that cities with stable boundaries were the most likely to lose population and employment.

Table 1.2 shows the size and regional distribution of the change in population and households for the periods 1960 to 1970 and 1970 to 1980. This table and the ones that follow present information by region, by annexation, and by our definition of old. The differences between the group of 50 annexing central city areas and the 35 areas with relatively static boundaries are clear. Between 1960 and 1970 the average increase in population for the annexing cities was 22.3%, compared to an average decline in population of 3.4% for cities with fixed boundaries. Similarily, for the 1970–1980 period annexing cities had population increases of 9.3%, whereas the population in cities with fixed boundaries declined 10.3%. Although the majority of these nonannexing winter cities are in the Northeast and upper Midwest, the ability to annex seems to play a significant role in the pattern of population change that will hold in any region.

Further support that the ability to annex is the principal distinction between growing and declining cities is provided in Table 1.2. The column labeled "Old" represents the group of central cities whose 1980 populations were less than their 1930 populations. When viewed in a regional context the relationship between old and fixed boundaries becomes clear. In the Northwest 14 of 18 cities had populations in 1980 that were less than the population in those cities in 1930. The cities in the Northeast were, with one exception, unable to annex any new area between 1960 and 1982. The one exception—Albany, New York—annexed less than 5% of its territory. The cities of the Northeast had population declines of 4.9% between 1960 and 1970 and 12.8% between 1970 and 1980. The average population declines, regardless of region, for nonannexing cities were 3.4% between 1960 and 1970 and 10.3% in the 1970–1980 period. When compared to a 22.3% increase in population between 1960 and 1970 and a 9.3% increase between 1970 and 1980 for annexing cities, the pervasive forces of decentralization and the role of annexation in population growth can be better appreciated.

There appear to be two distinct components of the Midwest: an older nonannexing group of cities and a newer annexing group. With

TABLE 1.1
Population and Land Area Changes with City Annexation and Age Characteristics (% change)

	Population			*Land Area*			*Fixed*	
	1960-1970	1970-1980		1960-1972	1972-1982		*Boundaries*	*Old*
	CC	OCC	CC	OCC	CC	CC		
Washington	-0.917	60.366	-15.608	10.65	0.00	0.0000	1	0
Springfield	2.431	20.874	10.320	5.62	0.00	0.0000	1	0
Paterson	0.714	18.372	-6.028	-4.46	0.00	0.0000	1	0
New York	1.452	26.271	-10.426	1.50	0.00	0.0000	1	0
Bridgeport	0.000	28.022	-8.333	3.00	0.00	0.0000	1	1
Hartford	-2.469	30.491	-13.924	19.16	0.00	0.0000	1	1
Baltimore	-3.621	34.722	-13.039	19.16	0.00	0.0000	1	1
Boston	-8.034	11.275	-12.168	-3.41	0.00	0.0000	1	1
Worcester	-3.297	17.483	-7.955	10.71	0.00	0.0000	1	1
Jersey City	-5.797	4.192	-13.846	-4.31	0.00	0.0000	1	1
Newark	-5.679	14.798	-13.874	-2.71	0.00	0.0000	1	1
Albany	-7.885	23.016	-12.062	10.11	4.30	1.9298	1	1
Buffalo	-13.158	14.470	-22.511	-0.11	0.00	0.0000	1	1
Rochester	-6.918	41.889	-18.243	9.39	0.00	0.0000	1	1
Syracuse	-8.796	26.513	-13.706	7.52	0.00	0.0000	1	1
Philadelphia	-2.697	22.607	-13.347	5.58	0.00	0.0000	1	1
Pittsburgh	-13.907	4.442	-18.462	-2.18	0.00	0.0000	1	1
Providence	-4.762	23.707	-7.353	5.05	0.00	0.0000	1	1
Mean Northeast	-4.9	23.5	-12.8	4.1	.2	.1	100	78

Fort Wayne	9.938	45.714	-2.825	19.61	46.40	8.6754	0	0
Gary	-5.172	34.667	-13.636	17.82	5.95	0.0000	0	0
Indianapolis	54.832	-17.045	-4.885	27.67	441.99	0.0000	0	0
Des Moines	-3.846	49.123	-4.500	31.76	0.64	4.2580	1	0
Wichita	8.661	-11.111	1.087	17.86	77.45	22.4862	0	0
Flint	-1.531	38.356	-17.098	19.14	13.09	-0.3659	0	0
Grand Rapids	11.299	20.070	-7.614	23.17	87.04	3.3408	0	0
Kansas City	6.737	20.908	-11.637	14.88	143.31	0.0000	0	0
Omaha	15.282	23.077	10.086	34.90	54.72	19.7590	0	0
Columbus OH	14.437	32.862	4.824	6.91	69.07	25.1988	0	0
Dayton	7.252	30.603	16.049	3.30	16.95	41.1978	0	0
Toledo	21.698	2.244	8.269	17.70	69.14	0.0594	0	0
Madison	37.302	23.158	1.734	30.77	44.29	8.0936	0	0
Milwaukee	3.239	27.747	11.297	10.93	5.56	0.0000	0	0
Chicago	5.183	24.045	10.695	23.70	0.36	0.0000	1	1
Detroit	9.521	28.489	20.318	5.65	0.00	0.0000	1	1
Minneapolis	6.784	56.058	13.612	18.24	0.00	0.0000	1	1
St. Louis	17.067	28.508	27.170	6.26	0.00	0.0000	1	1
Akron	5.172	27.937	13.818	4.96	0.35	8.1919	0	1
Cincinnati	9.960	21.830	14.823	9.01	1.42	0.0000	1	1
Cleveland	14.384	27.106	23.467	0.91	0.00	0.0000	1	1
Youngstown	9.735	17.730	16.176	8.13	3.30	0.0000	1	1
Mean Midwest	3.6	24.9	11.9	16.1	49.1	20.2	36	36
Birmingham	-11.765	15.263	-5.333	19.8	10.01	23.2707	0	0
Mobile	-2.062	10.714	5.263	30.11	-23.29	0.0000	1	0
Jacksonville	150.746	-88.976	13.294	7.14	2400.00	0.0000	0	0
Miami	14.777	44.946	3.892	37.23	0.00	0.0000	1	0

TABLE 1.1 (*continued*)

	Population			*Land Area*			*Fixed*	
	1960-1970	1970-1980		1960-1972	1972-1982		*Boundaries*	*Old*
	CC	OCC	CC	OCC	CC	CC		
Tampa	8.333	63.924	3.036	67.37	15.28	-0.0855	0	0
Atlanta	1.848	68.809	-14.315	41.43	2.73	-0.1826	1	0
Columbus GA	32.759	-16.832	9.740	-16.67	746.94	0.3460	0	0
Baton Rouge	8.553	54.545	109.697	-100.00	31.30	76.2666	0	0
New Orleans	-5.423	62.007	-6.071	39.16	0.00	0.0000	1	0
Shreveport	10.976	-4.274	13.187	13.39	77.21	29.9812	0	0
Jackson MS	6.250	38.158	32.680	12.38	39.78	65.1633	0	0
Charlotte	19.900	46.087	30.290	-4.76	18.90	90.9940	0	0
Oklahoma City	12.963	46.524	10.109	29.56	98.03	0.0099	0	0
Tulsa	26.820	-7.643	9.063	43.45	265.72	8.9471	0	0
Knoxville	56.757	-12.109	0.575	28.89	208.81	-0.1032	0	0
Memphis	25.352	-17.514	3.692	23.29	82.72	12.0350	0	0
Nashville	150.588	-75.850	6.808	100.000	1651.02	0.000	0	0
Austin	34.946	72.000	37.450	72.093	54.08	67.875	0	0
Corpus Christi	22.156	-18.367	13.725	17.500	350.80	0.600	0	0
Dallas	24.300	61.959	7.109	48.101	5.04	24.940	0	0
El Paso	16.667	0.000	31.988	48.649	7.02	96.587	0	0
Fort Worth	10.370	70.393	-2.036	47.554	60.00	9.870	0	0
Houston	31.343	58.750	29.383	69.423	37.60	30.134	0	0
San Antonio	11.414	63.281	20.031	19.617	33.85	35.190	0	0
Norfolk	0.000	65.190	-11.429	44.061	3.81	0.679	1	0

Richmond	13.699	24.074	-12.048	39.179	62.96	0.000	0	0
Louisville	-7.436	39.222	-17.452	15.269	5.26	1.883	0	1
Mean South	24.6	24.6	11.9	29.6	231.30	11.3	18.5	4
Phoenix	32.346	72.321	35.972	86.010	37.43	33.529	0	0
Tucson	23.585	69.231	26.336	128.409	18.57	27.269	0	0
Anaheim	31.845	135.663	23.928	41.411	28.42	19.770	0	0
Fresno	24.060	6.466	32.121	20.243	55.69	100.872	0	0
Los Angeles	12.398	19.969	4.885	7.597	2.53	0.245	1	0
Sacramento	32.984	26.097	8.268	35.165	108.68	2.812	0	0
San Bernadino	39.462	42.321	21.222	41.487	74.11	11.521	0	0
San Diego	21.466	44.009	25.862	49.168	65.59	-0.477	0	0
San Francisco	-8.644	31.948	-5.566	9.941	1.22	0.511	1	0
San Jose	118.137	41.324	43.146	6.300	163.51	14.769	0	0
Denver	4.260	63.678	-4.280	54.635	35.13	12.747	0	0
Honolulu	25.60		21.338		0.00	0.000	1	0
Albuquerque	20.896	18.033	36.626	22.222	49.27	26.386	0	0
Portland	2.688	39.421	-4.188	39.936	37.92	13.610	0	0
Salt Lake City	-7.407	47.674	-6.857	58.268	5.89	26.719	0	0
Seattle	-2.178	64.706	-6.164	26.071	23.25	0.000	0	0
Spokane	-6.077	20.833	0.588	47.414	0.00	8.780	0	0
Tacoma	4.762	47.977	3.247	27.734	1.46	0.000	1	0
Mean West	20.5	46.6	14.2	41.3	39.4	16.0	22	0

a. 1 = nonannexing fixed boundary city; 0 = annexing city.

b. 0 = city that gained population from 1930 to 1980; 1 = city that lost population from 1930 to 1980.

TABLE 1.2
Average Changes of Population—Households and Land Area by Region and Grouping

Variable	*Total*	*Fixed*	*Annex*	*Old*	*Not Old*	*North-east*	*Mid-west*	*South*	*West*
Percentage change population, central city									
1960-1970	12.1	-3.4	22.3	-7.5	16.6	-4.9	3.6	24.6	20.5
1970-1980	1.2	-10.3	9.3	-15.1	7.5	-12.8	-11.1	11.9	14.2
Outside city									
1960-1970	28.9	30.2	28.0	24.7	33.1	23.5	24.9	24.6	46.6
1970-1980	23.0	12.2	30.3	6.7	27.8	4.1	16.1	19.6	41.3
Percentage change households, central city									
1960-1970	18.3	2.1	29.5	-1.3	22.4	.2	9.6	31.9	26.4
1970-1980	15.8	2.7	24.9	-2.0	24.0	-.1	3.5	29.5	31.0
Outside city									
1960-1970	33.2	34.3	32.4	27.8	37.9	24.9	28.0	31.6	50.2
1970-1980	47.0	34.9	55.5	27.3	51.5	26.5	36.2	47.7	71.2
Percentage change, city land area									
1960-1972	94.6	.1	160.8	.7	98.8	.2	49.1	231.3	39.4
1972-1977	5.8	.1	9.7	0	8.2	0	2.4	12.4	5.8
1977-1982	5.2	0	8.7	.5	7.1	.1	3.5	6.9	9.6
City population 1980 as a percentage of 1930	276.6	125.0	382.7	79.9	353.0	86.2	126.2	346.5	545.8

the exception of Des Moines, which suffered population declines in each of the two decades following 1960, each of the eight cities in the Midwest that annexed less than 5% of its land area in either of the two decades between 1960 and 1980 had smaller populations in 1980 than they had in 1930. Although not an old city by our definition, Dayton, Ohio, has been a major population loser over the previous two decades, in spite of significant annexations. It is not the purpose of this chapter to determine whether annexations masked the effects of decentralization or actually changed the process, but there appears to be a relationship between population growth and annexation.

In contrast to the picture of the Northeast and older Midwest, the South is a region dominated by annexing central cities. Of the 27 central city areas in the South, 22 annexed at least 5% of areas in the periods between 1960 and 1982. The average increase in land area for the city areas in the South between 1960 and 1972 was 231.3%, a number somewhat affected by the consolidation of Jacksonville with the remainder of Duval County. Using measures of land areas standardized over time, the change in land area in the South was 12.4% for the five-year period between 1972 and 1977 and 6.9% between 1977 and 1982. The only city in the South that had fewer people in it in 1980 than in 1930 was Louisville, which was able to increase its land area by only a modest 5.26% between 1960 and 1982. The mean of the growth rates of population was 24.6% for the central cities of the South between 1960 and 1970. This growth rate slowed to 11.9% between 1970 and 1980, as the rate of annexation was reduced.

The West in many aspects is similar to the South, with annexation and growth commonplace. No city in the West had a lower population in 1980 than it had in 1930. Only 4 of the 18 central city areas in the West failed to increase their land areas in the 1960–1972, 1972–1977, or 1977–1982 periods. One of those places, Los Angeles, underwent considerable annexation prior to World War II. A second city, Honolulu, is a consolidated city-county.

The winter cities, whether in the North or not, are old cities, often in long-term decline, and their inability to annex has limited their ability to grow. The cities in the old group were able to increase their land area by only 1% between 1960 and 1980, and they lost approximately 21% of their population. Those new, growing, and annexing cities in the South, West, and Midwest more than doubled their land area and increased their populations by 25% in the same period. Because the process of decentralization and the regional redistribution of population have occurred simultaneously, it is difficult to assess a causal relationship.

Some sense of the scope of the redistribution of population among the regions within the United States can be obtained by looking at the change of the outside central city populations. During the 1960–1970 period the outside central city population of the nonannexing group grew at a rate that was only 2% greater than the growth of the annexing group, in part reflecting the deleterious effects of central city annexing on outside central city populations. Despite continued annexations in many central cities, the period between 1970 and 1980 was one in which the outside central city population of the group of annexing central cities increased by 30.2% compared to a 12.2% increase for the nonannexing places. There were declines in outside central city populations in a number of major metropolitan areas, apart from those caused by annexation.

The information presented in Table 1.2 suggests that the regional differences in outside central city population are not as great as the regional differences of central city population changes between 1960 and 1970. By the decade of the seventies OCC growth in the South and the West far outstripped that of the Northeast and Midwest, in spite of central city annexation that reduced outside area populations. Movements in population between central cities and their surrounding areas are not directly measured, so one can only surmise that the people of the winter central cities have migrated to their surrounding areas, but the resulting redistribution of income and households have affected the distribution of economic activity.

FORCES OF DECENTRALIZATION

Table 1.3 presents the level and changes in the level of total income, households, and population in cities and their surrounding areas. With the exception of the Midwest, the difference between the redistribution of population and the redistribution of households is mainly a difference of magnitude. When viewed on a regional basis, the number of households increased more rapidly than population in the places where population increased and decreased less rapidly than population in regions where population decreased. The number of households actually increased slightly in the Northeast between 1960 and 1970 in spite of an average population decline of nearly 5% for the period. Classifying cities by annexing activity shows similar results. Annexing cities increased 29.5% compared to a 2.1% increase in households for nonannexing places. The group of old places had the number of households decline by 1.3%, whereas new cities increased 22.4%. The relative attractiveness of central cities for small

TABLE 1.3
Changes in Current and Real Total Income, and Central City/Outside Central City Per Capita Income Ratios

Variable	*Total*	*Fixed*	*Annex*	*Old*	*Not Old*	*North-east*	*Mid-west*	*South*	*West*
Percentage change									
Current income, central city									
1960–1970	74.7	51.1	92.2	44.4	86.5	50.2	57.8	100.8	82.6
1970–1980	125.3	88.7	152.5	73.3	145.6	74.8	95.0	155.4	169.3
Outside city									
1960–1970	128.7	120.9	134.3	207.9	136.9	107.3	113.6	141.7	150.9
1970–1980	199.5	159.6	229.1	141.8	223.1	133.4	174.4	237.4	247.2
Real income, central city									
1960–1970	-13.2	-24.8	-4.4	-28.2	-7.2	-25.3	-21.5	-0.1	-9.2
1970-1980	-15.6	-29.3	-5.4	-35.1	-8.0	-34.5	-26.9	-4.3	0.9
Outside city									
1960–1970	13.4	9.6	16.2	3.1	17.5	2.8	5.9	19.9	24.5
1970-1980	12.2	-2.8	23.3	-9.6	21.1	-12.6	2.8	26.4	30.1
Ratio central city to outside city per capita income									
1960	105.1	95.8	111.5	90.8	110.3	89.8	98.3	114.7	114.1
1970	96.5	87.4	103.0	85.1	101.0	84.5	91.9	105.8	101.5
1980	89.9	79.8	97.0	77.0	94.9	75.9	84.4	99.9	96.8

households probably reflects the coming of age of the postwar baby boom generation interacting with the increasing prevalence of separation, divorce, single-parent, and single-person households.

This redistribution of population and households has yielded a redistribution of income. The one constant would appear to be that regardless of classification total income in central cities grew less rapidly than in outside areas. The increases between 1970 and 1980 in nominal total income ranged from 74.8% in the Northeast to 169.3% for central cities in the West; for outside central city areas the range was from 133.4% to 247% for the same regions. The change in income for the bounded group of nonannexers was 88.7%, whereas the annexing cities had increases in nominal total income of 152% for the 1970–1980 period. The changing relative positions of central city and outside central city areas can be seen from the ratio of central city to outside central city per capita income. Regardless of region, age, or annexing ability, per capita income in central cities has fallen when compared to per capita income in outside central city area counterparts. The group of nonannexing, old cities has fared the worst when compared to their outside areas, with central city income less than 80% of outside central city per capita income in each of these groups.

Although the relative changes in nominal total income shows us the deteriorating position of central cities vis-à-vis their surrounding areas, the true decline of central city income can best be seen when the change in real income is analyzed. Comparing the real income changes to nominal income changes shows that the decline in real income between 1970 and 1980 was greater than the decline between 1960 and 1970, which was the period of greater decline in nominal terms. The real income changes in Table 1.3 show unambiguously that regardless of region, regardless of boundary changes, the relative position of central city income fell with respect to outside central income. Regional differences in changes in real income between 1970 and 1980 in part reflect the economic dislocations that have resulted from changing energy prices, slumping domestic automobile production, and the displacement of the American steel industry. The greatest declines in real total income in central cities were in the old, nonannexing winter cities of the Northeast and the older Midwest.

The changes in real income detailed previously suggest that central city areas in these regions have continued to be centers for the low-income population. Although the relative attractiveness of central cities for households has been explained in terms of the new location for young upwardly mobile individuals, the hard evidence suggests that at least in the old cities with fixed boundaries the effects of these individuals are more than offset by low-income households. In the old cities of the Northeast and Midwest real total income declined between 1970 and 1980. The accompanying shift in income from central cities to outside central city areas can be translated into a change in buying power. Those economic activities most closely linked to income, such as retailing functions, have followed the redistribution of income and buying power out of the central cities.

The change in retail activity must be put in the context of the change in total economic activity in the relevant market area—generally, but not always, the metropolitan area. Thus, the change in central city retail activity presented in Table 1.4 is shown in relation to the change in total employment activity. The decentralization of retail activity seems to correspond to the change in the proportion of the SMSA population found in the central city. Between 1967 and 1982 central city retail employment as a percentage of SMSA employment fell in each region, with the central cities of the Northeast losing the greatest proportion of retail employment. In 1967 the fraction of SMSA retail employment in the fixed boundary central cities was 52.2%; by 1982 that fraction had fallen to 35.9%. In contrast, those central cities that were active annexers accounted for 72.3% of the retail employment in their SMSAs in 1967; by 1982 even the annexing cities found their role diminished to 60.7%.

Part of the explanation for these changes can be found in the changing fractions of SMSA population in central cities presented in Table 1.4, but part of the change is due to the changing shares of real income—buying power—in the central cities and their surrounding areas. Viewing the long-term trend of the change in central city retail trade employment, one can see that for the cities of the Northeast and Midwest retail employment declined significantly during the period from 1967 to 1982. Our group of winter cities—that is, those cities in static boundaries and those cities classified as old—had the greatest declines. Those central cities that actively annexed their surrounding areas had an increase in retail trade employment of 44% in the 15-year period. The nonannexing cities had retail trade employment declines averaging 11%, along with 10.3% population declines between 1970 and 1980. Even though the central cities of the South and West had increases in retail employment, they were far outstripped by the growth of employment in their surrounding areas. Thus, the decentralization of retail employment, like the decentralization of total income, in regional terms varies only by degree.

The change of total employment, however, has distinctive regional characteristics associated with it. Two measures of employment are presented in Table 1.5. The first is based on the information provided in the various economic censuses, and only includes employment in

TABLE 1.4
The Decentralization of Population and Retail Trade Employment, 1967–1982

Variable	*Total*	*Fixed*	*Annex*	*Old*	*Not Old*	*North-east*	*Mid-west*	*South*	*West*
City retail employment as percentage of SMSA									
1967	64.0	52.2	72.3	46.8	70.1	44.9	61.6	78.5	64.3
1972	57.5	44.2	66.9	37.6	64.3	37.9	54.3	73.7	54.0
1977	53.1	38.5	63.4	32.2	60.2	33.0	48.6	70.1	53.3
1982	50.4	35.9	60.7	28.9	57.8	30.9	45.6	67.3	51.3
City population as percentage of SMSA									
1960	52.0	45.9	56.1	42.7	55.8	41.7	52.5	59.8	49.7
1970	49.0	39.0	55.6	35.8	53.1	35.9	48.2	60.6	45.0
1980	44.6	34.0	51.9	30.9	49.2	32.2	42.1	57.7	40.4

manufacturing, retail trade, wholesale trade, and a group of selected services. The data presented here have been assembled in metropolitan areas defined in 1970 terms, using employment classifications that are held constant to the greatest degree possible. The second definition of total employment is from the Census of Population, and includes all civilian employment in metropolitan areas generated on the basis of a place of work. Combining these two sources of information gives us some insight into the regional redistribution of employment and the decentralization of economic activity.

The decline of employment in the metropolitan areas of the Midwest in part reflects the economic recession of 1982. The lower rate of growth of total employment on metropolitan areas having central cities with fixed boundaries suggests that there are a multitude of forces at work in the change of central city employment. Turning to the Census of Population data for total employment, we see that the general decline of central city employment that occurred in the Northeast and Midwest during the 1960–1970 period was mitigated in the Northeast and reversed in the Midwest in the decade of the 1970s.

In the period ending in 1980, only the central cities of the Northeast declined in total employment. Even more important, the nonannexing central cities, which had employment declines of 7.7% during the 1960–1970 period, had employment gains of 6.5% between 1970 and 1980. The employment increases for annexing central cities were 19.7% and 36.5% for the same periods. The group of old cities had a substantial reduction in the rate of decline, from −12.2% to −1.4% in the two decades.

RESURGENCE OF CENTRAL CITY EMPLOYMENT

The increasing importance of the central city as a center of employment in relation to their populations is presented in Tables 1.6 and 1.7. Between 1960 and 1970 the ratio of employment to population declined in all regions but the West. Employment/population ratios increased from 15% to 27% between 1970 and 1980, in part reflecting national changes in labor force participation rates from which cities benefited. Even in cities with fixed boundaries, which have been the most adversely affected by any other measure of change, the ratio of employment to population increased by 18.5% between 1970 and 1980. Central city unemployment rates, which appear in Table 1.7, also hold some hope for the older winter cities.

TABLE 1.5
Selected Measures of Employment Change for Central City, Outside Central City, and SMSA

Variable	*Total*	*Fixed*	*Annex*	*Old*	*Not Old*	*North-east*	*Mid-west*	*South*	*West*
Percentage change in total employment census of population place of work)									
Central city									
1960–1970	8.4	-7.7	19.7	-12.2	15.0	-10.4	-0.6	19.1	22.4
1970–1980	24.2	6.5	36.5	-1.4	33.8	-0.1	8.3	39.9	44.3
Outside city									
1960–1970	30.4	38.1	25.0	31.3	31.9	26.1	32.1	26.2	38.9
1970–1980	47.9	39.7	53.7	33.6	53.0	28.7	36.5	55.4	69.9
SMSA									
1960–1970	15.4	8.4	20.3	3.9	19.7	4.2	9.7	20.1	26.6
1970–1980	34.5	21.1	43.8	15.7	41.1	13.8	20.4	45.7	55.5
Selected employment totals, economic censuses (SMSA)									
1967–1972	13.8	6.8	18.2	1.4	18.6	1.4	6.0	23.9	20.2
1972–1977	12.9	4.9	18.4	1.5	17.3	0.4	8.0	16.4	26.1
1977–1982	12.4	5.6	17.1	0.2	17.1	5.1	-2.2	20.0	26.1
1967–1982	48.5	20.1	68.3	3.4	65.2	8.0	12.5	75.4	92.4

TABLE 1.6
Central City Employment to Population Ratios and Rates of Unemployment for Selected Years

Variable	*Total*	*Fixed*	*Annex*	*Old*	*Not Old*	*North-east*	*Mid-west*	*South*	*West*
Ratio of the percentage change in employment to the percentage change in population									
Central city									
1960–1970	98.7	95.8	100.2	95.1	100.1	94.4	97.5	99.2	103.7
1970–1980	122.7	118.5	125.7	116.2	125.2	114.9	120.8	126.9	126.8
Outside city									
1960–1970	100.5	105.0	97.3	104.5	98.9	101.2	105.6	98.6	96.1
1970–1980	121.9	124.5	120.1	125.1	120.7	123.5	120.5	120.9	123.6
Central city unemployment rates									
1975	7.4	8.0	7.0	8.3	7.2	7.4	9.2	6.6	6.6
1982	10.8	11.6	10.2	12.4	10.2	11.4	13.8	8.6	10.0
1984 (Sept.)	9.3	10.9	8.2	11.3	8.6	11.6	9.7	7.2	9.6

TABLE 1.7
Changes in the Level of Employment and Unemployment Rates for Selected Years
(Percentage change in employment)

	Total Employment Decennial Census, Place of Work Basis					*Unemployment Rates Central Cities*		
	1960-1970		*1970-1980*		*Selected Categories SMSA*			
	CC	OCC	CC	OCC	1967-1982	1984	1982	1975
Washington	0.185	84.80	13.678	57.40	22.424	8.3	10.6	7.6
Springfield	-4.132	46.22	1.724	1.72	-2.500	4.9	10.1	12.8
Paterson	-7.477	25.90	2.020	9.34	11.006	10.9	15.9	17.1
New York	-11.814	13.31	-7.233	47.77	-4.574	7.9	9.6	10.6
Bridgeport	-10.000	25.09	-13.889	28.66	-9.735	6.5	10.4	13.5
Hartford	-18.333	17.52	8.163	52.61	-2.793	6.4	9.1	11.0
Baltimore	-7.059	52.05	5.570	23.91	3.866	7.8	11.4	10.1
Boston	-13.578	10.60	5.237	30.61	3.949	3.8	9.1	12.9
Worcester	-7.609	13.21	7.059	47.44	3.704	4.0	10.2	11.8
Jersey City	-35.417	-29.75	-12.903	-12.90	-5.096	10.5	15.3	12.1
Newark	-33.929	-8.52	-21.622	38.86	2.461	10.3	15.4	17.7
Albany	7.586	26.80	8.974	28.03	3.125	6.0	7.6	7.5
Buffalo	-17.557	28.24	-9.722	10.80	-1.712	11.8	16.9	14.7
Rochester	-4.390	89.10	2.551	59.87	-0.317	7.6	10.1	9.3
Syracuse	-12.712	35.83	6.796	20.59	3.252	6.7	9.3	9.3
Philadelphia	-9.032	25.96	-11.702	43.48	-1.707	6.9	9.0	9.7

Pittsburgh	2.034	-2.33	5.648	5.65	-2.419	8.1	10.2	9.3
Providence	-4.571	16.69	8.383	22.21	2.817	4.8	14.3	12.2
Mean Northeast	-10.4	-.6	-0.1	22.40	8.0	7.4	11.4	11.6
Fort Wayne	3.750	37.13	18.072	12.02	10.059	7.0	13.5	11.0
Gary	-2.158	45.18	-3.676	4.63	2.740	21.2	24.4	10.6
Indianapolis	12.692	-9.20	29.352	-11.78	1.240	6.3	10.2	6.8
Des Moines	3.226	19.81	22.917	110.85	16.129	5.5	9.4	4.7
Wichita	15.730	-19.08	38.835	2.30	-16.495	6.4	9.0	5.6
Flint	0.000	31.25	7.921	44.74	-3.305	14.6	27.0	15.3
Grand Rapids	4.494	4.49	13.978	39.72	7.143	10.3	14.9	11.2
Kansas City	5.639	30.02	5.694	45.82	6.015	5.5	10.4	8.9
Omaha	11.111	29.63	28.667	16.25	17.778	5.1	7.9	5.5
Columbus OH	12.037	36.66	29.339	55.47	27.108	7.1	10.0	7.9
Dayton	3.185	16.46	-17.901	38.87	2.577	10.5	17.0	10.9
Toledo	16.429	11.45	1.840	31.43	14.815	9.3	13.0	10.5
Madison	33.333	26.12	36.364	52.17	20.930	4.3	6.1	5.0
Milwaukee	-12.640	69.46	3.215	31.45	1.719	7.2	13.1	10.0
Chicago	-16.030	60.31	-4.618	48.37	1.137	8.6	11.7	9.0
Detroit	-23.234	15.62	-20.354	31.98	-0.423	13.3	20.3	17.4
Minneapolis	2.299	106.26	7.294	104.83	7.711	4.5	6.5	7.8
St. Louis	-17.913	49.49	-16.547	29.82	-2.930	12.6	15.7	11.5
Akron	-18.248	19.72	1.786	40.43	1.948	10.8	14.2	10.4
Cincinnati	-11.151	28.28	6.073	29.58	5.842	9.2	13.2	10.6
Cleveland	-17.913	49.49	-16.547	29.82	-2.930	12.6	15.7	11.5
Youngstown	-14.563	50.44	4.545	8.85	10.656	16.9	24.4	13.3
Mean Midwest	-0.6	32.1	8.3	36.5	12.5	9.2	13.8	9.7
Birmingham	-0.641	17.971	19.355	46.68	11.364	10.9	16.9	7.7
Mobile	-8.235	-3.71	37.179	57.65	20.833	12.3	14.3	5.8

TABLE 1.7 (*continued*)

	Total Employment Decennial Census, Place of Work Basis				*Unemployment Rates Central Cities*			
	1960-1970		*1970-1980*		*Selected Categories SMSA*			
	CC	OCC	CC	OCC	1967-1982	1984	1982	1975
Jacksonville	76.923	-100.00	21.256	0.00	34.177	5.6	6.9	6.8
Miami	-2.551	64.48	70.681	51.06	36.744	9.0	12.8	12.9
Tampa	18.391	69.31	42.718	132.86	39.286	5.8	8.5	9.3
Atlanta	9.155	108.39	12.581	124.50	22.535	6.3	8.9	10.3
Columbus GA	4.082	32.68	19.608	-13.62	17.647	7.1	8.1	7.4
Baton Rouge	2.857	26.67	77.778	-100.00	25.000	7.0	7.8	5.7
New Orleans	-4.032	77.82	11.765	39.08	11.585	9.0	9.2	8.5
Shreveport	-4.225	0.49	50.000	72.39	30.769	7.4	9.7	7.1
Jackson MS	20.000	6.79	43.056	51.79	26.316	6.6	7.6	5.7
Charlotte	23.810	39.68	30.000	30.00	23.913	4.6	6.0	8.0
Oklahoma City	25.806	4.63	35.897	148.29	32.609	5.6	4.7	8.1
Tulsa	17.500	-24.46	53.901	139.40	16.092	7.0	6.0	6.4
Knoxville	27.143	-16.32	31.461	76.79	12.329	6.7	9.7	6.6
Memphis	12.136	4.63	30.736	30.74	18.382	7.1	10.4	7.2
Nashville	88.776	-42.92	34.595	133.19	17.757	4.1	7.1	6.8
Austin	38.667	54.24	101.923	22.97	47.059	3.2	5.0	4.1
Corpus Christi	11.667	11.66	53.731	-4.64	27.273	7.3	7.3	6.4
Dallas	33.634	63.74	50.562	135.67	19.749	3.8	6.2	4.9
El Paso	9.091	9.09	73.958	-43.09	29.167	9.9	10.9	7.9

Fort Worth	-2.532	103.197	45.455	57.903	8.276	4.2	8.3	6.7
Houston	47.059	55.472	52.522	78.638	34.877	6.4	7.0	5.3
San Antonio	45.143	-24.221	44.882	35.583	29.897	4.8	6.8	7.7
Norfolk	14.208	73.426	0.957	77.877	28.571	4.5	7.0	6.7
Richmond	8.759	49.956	22.819	70.057	14.019	4.5	7.2	5.5
Louisville	5.650	44.128	8.556	44.915	10.270	6.3	10.9	8.5
Mean South	19.1	26.2	39.9	55.4	75.4	6.6	8.6	7.2
Phoenix	49.367	63.750	68.220	118.361	41.259	3.2	8.4	13.0
Tucson	54.839	-30.992	72.917	122.181	44.444	3.8	9.6	9.5
Anaheim	82.143	40.747	59.477	125.926	26.872	4.3	8.2	8.1
Fresno	14.516	1.463	64.789	51.874	23.077	11.3	12.1	8.3
Los Angeles	1.266	28.972	17.500	27.288	2.709	8.0	10.4	10.9
Sacramento	47.706	-5.427	24.074	85.619	11.111	8.1	12.6	9.4
San Bernadino	14.706	19.827	42.735	56.044	26.829	8.3	13.9	5.7
San Diego	16.187	65.341	47.988	60.851	27.848	5.4	9.3	12.1
San Francisco	0.167	27.685	10.982	46.948	7.048	5.1	8.4	12.1
San Jose	40.000	72.957	65.873	81.119	23.618	5.1	8.8	10.3
Denver	16.599	73.943	37.153	123.277	36.923	4.8	7.0	8.1
Honolulu	19.858	50.000	47.337	33.333	33.333	5.3	6.1	7.7
Albuquerque	19.737	28.326	83.516	47.533	-9.091	6.0	7.9	9.4
Portland	8.543	64.929	35.648	47.023	17.778	8.6	10.7	10.5
Salt Lake City	12.500	73.114	36.752	74.321	27.160	6.2	8.7	7.3
Seattle	12.252	63.421	24.779	82.075	-10.239	7.4	10.9	9.3
Spokane	6.154	1.346	26.087	57.863	21.053	8.3	12.6	10.1
Tacoma	-12.500	60.661	31.746	16.809	12.766	10.1	13.5	11.5
Mean West	22.4	38.9	44.3	69.9	92.4	6.6	10.0	9.6

The regional means of central city unemployment rates were much larger in 1975 than they were in 1984. The same is true for the difference between annexing and nonannexing cities, although the latter still possess higher unemployment rates. The unemployment rates of the winter cities are obscured by the mixed nature of unemployment. Automobile and steel production, which have been associated with the highest rates of unemployment, have been centered in the Midwest and Great Lakes area. Thus, as a group, winter cities have had high unemployment rates due to their industry mix. However, unemployment rates in the traditional winter cities of New England have been among the lowest in the nation.

Inasmuch as it was the decline of New England that triggered the concern for the winter cities, perhaps the decline of unemployment and the possible resurgence of employment in these areas will spill into other winter cities. Before the resurgence of the winter city is proclaimed, the growth of total employment in cities with fixed boundaries between 1970 and 1980—which was 6.5%—should be compared to the 36.5% increase in employment that occurred in the annexing cities. Additionally, the group of old cities had employment declines of 1.4%. Although much of this no doubt is associated with the uneven burden of the 1982 recession, the recent resurgence of New England is unique among the group of winter cities. When the recent declines in metropolitan population are considered, it seems that a reduction in the rate of decline may be no small feat.

The future of the winter city is far from certain. The main group of cities that has broken the pattern of decline, based on unemployment rates, is in New England. It is here that the post-World War II concept of decline initially surfaced, and where the debate about deindustrialization has been focused. The decline in areas specializing in steel, coal, and automobile production has not been limited to American cities, with many European cities having the same pattern of employment declines and high unemployment rates. If the long-term manufacturing decline of New England put cities there in a better position to respond to business cycles, then perhaps the future of some of the remaining winter cities will include a more vital economic base after undergoing a long-term decline.

The relative attractiveness of central cities in general, as centers of employment, and the widespread growth of households might be a positive sign; but one should still note that the negative signs far outweigh the positive for the old, nonannexing cities that dominate our group of winter cities.

Part I

The Condition of Winter Cities

□FOR SOME PEOPLE Miami or Palm Springs are their winter cities. But the winter propensities of so-called sun tourists are not the subject of our concerns. The concern for better conditions for winter cities, cities that experience a winter climate of discomfort and inconvenience, can be disaggregated to wintertime concerns for several different kinds of communities that include:

(1) ice-bound resource communities or mining towns above or close to the Arctic Circle;
(2) snow-bound cities with an average of 40 to over 120 snow-cover days, which approximates to the 40° latitude in the United States with a dip southward to include Taos, New Mexico, and the southern tip of the Rockies;
(3) cold-rain cities that experience little snow, but have substantial winter wetness such as Reykjavik, Iceland, and Oslo, Norway;
(4) ski-oriented resort communities such as Aspen, Colorado, and Lausanne, Switzerland, which are gifted with a dry and snow-rich environment that is steep in an appropriate number of places.

In addition to these four categorical distinctions, an additional concern must be expressed about the *severity factor*. A January average temperature belt of 0°F to 30°F excludes much of the United States, except for the Northeast and Midwest, and all of Europe west of Poland, which is warmed continuously by the Gulf Stream through the year. Further north a 0°F to −30°F belt begins slightly below Hudson Bay and includes most of Siberia. There the experience of winter resembles that of the Arctic Circle and requires extraordinary precautions to preserve and maintain life. It is said that in Siberia it is a capital offense to turn off the motor of a military vehicle between November and April. It is simply too cold for it to reignite.

Another aspect of winter in many American cities is its "unreliability." It may begin in an irregular fashion sometime in late October or November and may end well before the first day of spring or it can surprise us with a wet snowfall in April.

The design of and planning for livable winter cities represents, however, a challenge that can unite winter cities regardless of how the experience of winter may vary. The challenge is to combine the functional and practical aspects of the livable winter city with components that make the city both comfortable and interesting for its citizens. It requires a concern for both the structural and psychological aspects of urban environments.

THE SAD SEASON AND WINTER CARNIVALS

For many people the late dawns and early twilights of winter bring on mild depression and a mood of listlessness. But for some a more serious and severe depression occurs. Scientists call this a Seasonal Affective Disorder (SAD). SAD victims sleep fitfully, lose their energy and libido, and may binge on carbohydrates.

Recent research on the biological and medical effects of light has now enabled psychiatrists to experiment with light therapy. The use of both bright lights and bright colors—wearing red in February—can be helpful to individuals and to entire communities.

Midwinter festivals, which are common in Canadian and Northern European cities, can be replicated in other communities. Akron now has a "Chili" golf tournament during the first weekend of February. It attracts over 500 participants who play a round of golf on a frozen lake and the proceeds go to charity.

Whether called "winterfest," "winterlude," or "winterskol," winter carnivals have a somewhat indistinct relationship to pre-Lenten carnivals in more Catholic and southern countries. Winter sports, parades, and costumes and good fun outside are key elements of community-run festivals. The longest-running winter festival in North America appears to be in Anchorage and dates back to fur trappers in the 1870s. Now called the Anchorage Fur Rendezvous (the "Rondy"), this festival consists of more than 150 events over 10 days that include a dogsled race, a snowmobile race, a state fur auction, and an Eskimo blanket toss.

In Canada the oldest carnival is the Carnival de Quebec held in Quebec city over two weekends in February. More recently Winterlude in Ottawa was said to have drawn more than 450,000 participants in 1984. It consists of a number of events including a ski marathon, skate races on the Rideau Canal, snow sculptures, and sleigh rides.

Even universities are developing midwinter festivals. At the University of Chicago there is a mid-February event called by the Eskimo word, *Kuriasungnerk*, which is a combination of recreational and cultural activities.

These and other community-based efforts all seem to grow in popularity as part of local winter traditions, and reveal how attractive a midwinter outlet can be. The new popularity of cross-country skiing in many northern cities also provides new uses for municipal parks and golf courses.

DESIGN PERSPECTIVES FOR WINTER CITIES

A more significant urban planning approach concerns innovations in design and building materials, beginning with domed stadiums and extending to enclosed and heated bus shelters. Retractable glass panels over sidewalks can eliminate much discomfort to pedestrians in the wintertime.

As developed in cities such as Minneapolis, Calgary, Toronto, and Montreal, a new wide range of pedestrian sheltering devices such as tunnels, atria, gallerias, and skywalks now serve as demonstration models for other urban places. These forms of protection abound in downtown Melbourne, Australia, where there is no snow or frost but plenty of cold rain.

John C. Royle, the founder of the Livable Winter City Association in Canada, has written that "the dream winter city, lively, colorful and entertaining, must of necessity be one of fairly high density" (Pressman, 1985). The encouragement of greater mixed-use activities that incorporate a climate-protection principle represents a new challenge to urban designers and counters some of the forces contributing to urban sprawl.

Many of these new design initiatives in northern cities have been made possible because of a technical breakthrough by the Pilkington company in England, called the "Float-grass" process. New forms of transparent membranes also expand the prospects for enclosing large areas that can be protected from the harshness of winter weather.

Arni Fullerton, an urban planner in Edmonton, has proposed 14- and 35-acre transparent domes or tent structures that help to serve as "sun catchers" for consideration by several northern Alberta communities.

These new approaches to winter urban architecture recall the earlier efforts of Ralph Erskine.

THE WINTER ARCHITECTURE OF RALPH ERSKINE

An extremely important source of inspired thinking about urban design in cold climates has been the British architect Ralph Erskine. As a young architect committed to new ideas of urban planning and socialist development, Erskine set off for Sweden in early 1939, several months before the start of World War II.

Partially because many of his Quaker colleagues were being imprisoned in England as conscientious objectors, Erskine stayed in Sweden and struggled to start his career as an architect whose style one writer (Collymore, 1982) has described as "romantic functionalism."

Erskine's architecture is based upon two fundamental precepts:

> The buildings must be related to the climate and to the people who will inhabit and use them.

One of the most interesting examples of Erskine's architectural and planning style actually occurred in North America, where he helped to plan a new township at Resolute Bay in Arctic Canada that was established for both Eskimos and white southern Canadians. This design problem featured two different types of Canadians who had to live together in an extreme climate. This 1973 effort in the Northwest Territories of Canada drew not only upon Erskine's participatory Quaker style, but also upon his mid-1950s work for a distinctly Arctic architecture.

By 1958, Erskine had produced a drawing for an "ideal" town in the Arctic region of Sweden. Located on a south-facing slope, it was enclosed on the three other sides by a continuous, elongated building with a few scattered windows. Like the medieval walled cities of Europe, the plan was designed to keep out the "barbarians"—Arctic winds and blizzards. Erskine's windscreen building is a distinct innovation.

Within the shelter of the city-walled building is created a "sun trap" within which individual homes and other facilities are located on a pattern of medium to high density. By 1964, Erskine had completed a housing scheme for a company mining town above the Swedish Arctic Circle at Svappavaara for the state-owned L.K.A.B. iron ore excavation company that included his first actual windscreen building, a kinked terrace along the north side of the town.

Although much of Erskine's Arctic architecture is designed for relatively small and isolated "resource communities" that are reminiscent of moon station proposals, he also proposed what is called a "grammar for high altitudes." It includes concerns for (1) cold air drainage, (2) internally sloping roofs with special drain pipes that retain snow as an insulation aid, (3) special windows that shade against the continuous sun in summer and trap and reflect sparse sunlight in the wintertime, (4) rounded corners that reduce drifting of snow, (5) solar radiation devices, (6) roofing umbrellas that retard the formation of dangerous icicles because of separation from the warmth of the housing shell, and (7) an appreciation of the visual value of snow and other winter aesthetics.

These elements of urban housing design are especially applicable in any environment where snow, winter wind, and sunlight are more than just a temporary January inconvenience. As Collymore (1982: 30) writes:

> By carefully considering those factors of temperature, wind direction, seasonal change, special and more ordinary climatological conditions, he has developed an adaptable architecture suitable for varied conditions. His architecture is the antithesis of that common twentieth century kind where the same building is erected worldwide and the different climates held at bay by expensive energy-guzzling machinery.

At the same time, however, Erskine has also rejected the fantasy of winter towns under some kind of massive geodesic dome or membrane roof. He has written:

> Although the Arctic and sub-Arctic winter has many disadvantages, the crisp clean air of spring, the brilliant sun on the snow, the northern lights, are all appreciated as much by Northerners as by Southerners who spend money on winter sport holidays. Instead of the science fiction city under a dome, other, more subtle building forms should be devised with sheltered outdoor walkways open to the sky, the sun and to falling snow, interlocked with a system of enclosed, heated and

> daylit streets for bad weather, with a third circulation system of car routes covered where possible against snow drifting and perhaps running in building basements or on the northern side of the structures. Apart from the protective wall buildings mentioned previously, belts of trees and bushes can be planted to provide windbreaks and snow traps.

The architectural perspective about winter cities is an important element in the livable winter city movement in Canada, and we need to develop a stronger perspective about the dynamics of Canadian urban development.

URBAN CANADA: AN INITIAL PERSPECTIVE

Canada is a country of about 26 million people and has over 30 cities with populations of over 100,000. The 26 largest cities are shown in Table I.1.

The livability of Canadian cities is for most urban students and scholars symbolized by Toronto, which in 1976 was described by *Time* magazine as the world's most livable city (Time, Canada edition, June 20, 1976). About the same time *Fortune* called it "the world's newest great city," and *Harper's* called it "a model of the alternative future."

More recently, as Toronto prepared itself for its participation in the World Series, a Hollywood producer was shooting a film in the streets of Toronto. Because Toronto is so clean, the director ordered that a load of garbage be spread around to make the street scene seem more authentic. The ultimate irony was that during a cast coffee break, the street was cleaned up again (New York Times, October 6, 1986)!

Although Montreal, from the time of its Expo in 1967 through the Olympics in 1976, seemed destined to be the primate city of Canada, the emergence of a separate political party in the mid 1970s caused a certain amount of corporate flight. Now, in the words of my oldest son, Montreal resembles "a French Philadelphia" whereas Toronto is more what Chicago would look like if there wasn't a New York City but only a Boston.

Vancouver is the third jewel in the Canadian system of cities. Its dramatic vista on the Pacific Coast will become more familiar following its 1986 Expo.

TABLE I.1
Canada's Largest Cities

City	Population
Montreal, Quebec	1,020,900
Calgary, Alberta	620,700
Toronto, Ontario	614,760
Winnipeg	588,570
Edmonton, Alberta	560,000
North York, Ontario	559,520
Scarborough, Ontario	443,350
Mississauga	325,850
Hamilton, Ontario	308,100
Ottawa, Quebec	302,480
Etobicoke, Ontario	298,700
London, Ontario	266,300
Windsor, Ontario	192,380
Quebec City, Quebec	176,600
Regina	162,600
Saskatoon	154,210
Brampton, Ontario	150,000
Surrey, British Columbia	147,130
Kitchener, Ontario	141,430
Burnaby, British Columbia	136,500
St. Catherines	136,500
Oshawa, Ontario	118,845
Burlington, Ontario	114,850
Halifax, Nova Scotia	114,590
Thunder Bay, Ontario	112,480

But urban Canada consists not only of Toronto, Montreal, and Vancouver but of four to five other elements including:

- the maritime cities such as Halifax and St. Johns;
- the prairie cities of Winnipeg, Calgary, and Edmonton;
- the manufacturing cities around the Great Lakes including Windsor and running up through Hamilton and Kingston up to the mouth of the St. Lawrence; and
- primary-producing communities dependent upon agriculture, mining, or timber.

A Canadian geographer has described the main characteristics of the Canadian urban system as "a highly specialized economic base, dependent upon natural and economic forces beyond our control, with the urban nodes linked together in a hierarchic structure" (Simmons, 1976).

He goes on to indicate that "the prosperity of the majority of cities in Canada is based on the production of a single primary commodity." Besides the strong relationship of most Canadian cities to the resources in their immediate winterland, they can also be characterized as "highly integrated in the sense that changes in the economic base of one city are quickly communicated to other nearby cities and translated into relative rates of growth and decline."

There is also a strong sense of hierarchy with each small urban center linked by migration, communications, and transportation to the next highest-order place, and sometimes the competitive relationships between the highest-order cities (Toronto and Montreal, Winnipeg and Calgary) shift according to changing patterns of resource relationships (oil in Alberta) or the range of services offered (new cultural and recreational services in Toronto, including the Blue Jays).

Inasmuch as all Canadian cities have, in a very real sense, *more* winter than other North American cities, the success of the livable winter city movement in Canada is understandable. But some other differences should also be noted. In a forthcoming book Mercer and Goldberg (1986) challenge what has been called the myth of the North American city.

In their examinations of 277 SMSAs in the United States and 40 CMAs in Canada, they have detailed some of the distinctions between Canadian and American urban forms. At the Metropolitan scale, less than two-thirds (57%, 1971) of the housing stock comprised single-family units in Canada, whereas in the United States almost three-quarters (73%, 1970) of the metropolitan housing stock was of a single-family nature.

Other data indicate that the Canadian cities are "more compact . . . more readily served by mass transit and with a comparatively weakly developed freeway system" (Edmonton, 1985). Despite rapid decentralization in both countries since the end of World War II, Canadian cities have mean central densities twice as great as their United States counterparts. It is likely that the more densely settled urban areas in Canada contribute to the commercial success of such efforts as the Eaton Centre in Toronto and the West Edmonton mall, and create a greater sense of urbanization in all Canadian cities.

Goldberg and Mercer also comment on the lack of a spatially concentrated underclass in Canadian cities and their more healthy and stable fiscal condition based upon a greater proportion of the metropolitan area population being contained within the central city.

Before closing this section, mention must also be made about (1) the short-lived federal Ministry of Urban Affairs (MSUA), (2) the sub-Arctic community of Fermont, and (3) the Canadian Climate Impacts Program.

According to one involved participant the Ministry of Urban Affairs "was an innovation for the purpose of developing policy and advising government with respect to a broad policy field cutting across several departmental jurisdictions, and in a way that was deliberately divorced from the distractions of program delivery" (Gertler, 1985). The MSUA was an initiative of the Trudeau government in 1971 that lasted until it was disbanded in 1979. Although it successfully supported valuable urban studies policy research (Gertler, 1979), it could not overcome opposition from provincial politicians who argued that the Canadian Constitution gave them exclusive reign over municipal concerns.

Fermont is a new town in northern Quebec that is described by its architect as a fifth-generation sub-Arctic settlement (Schoenauer, 1976). Its basic planning concept follows Erskine's "windscreen building principle" and elaborates it further. Fermont's windscreen building is continuous and has "a pluri-use character embracing residential, commercial, recreational, educational and institutional facilities." A climate-controlled pedestrian corridor and mall links the ground floor of the five-story windscreen building to other facilities situated on the windward side of the building, which is designed to concentrate snow accumulation on its leeward side. Figure I.1 shows the site plan for the community.

The planning for future communities such as Fermont may be facilitated by the new Climate Impacts Program in Canada (CCP). The CCP is designed to improve the understanding of the socioeconomic consequences of climate. Its immediate priorities are (a) better techniques for climate impact assessment, (b) food production in northern areas, (c) water resources in the Great Lakes Basin, and (d) cooperation between all levels of government, universities, and the private sectors interested in the use of climate information in socioeconomic and resource management.

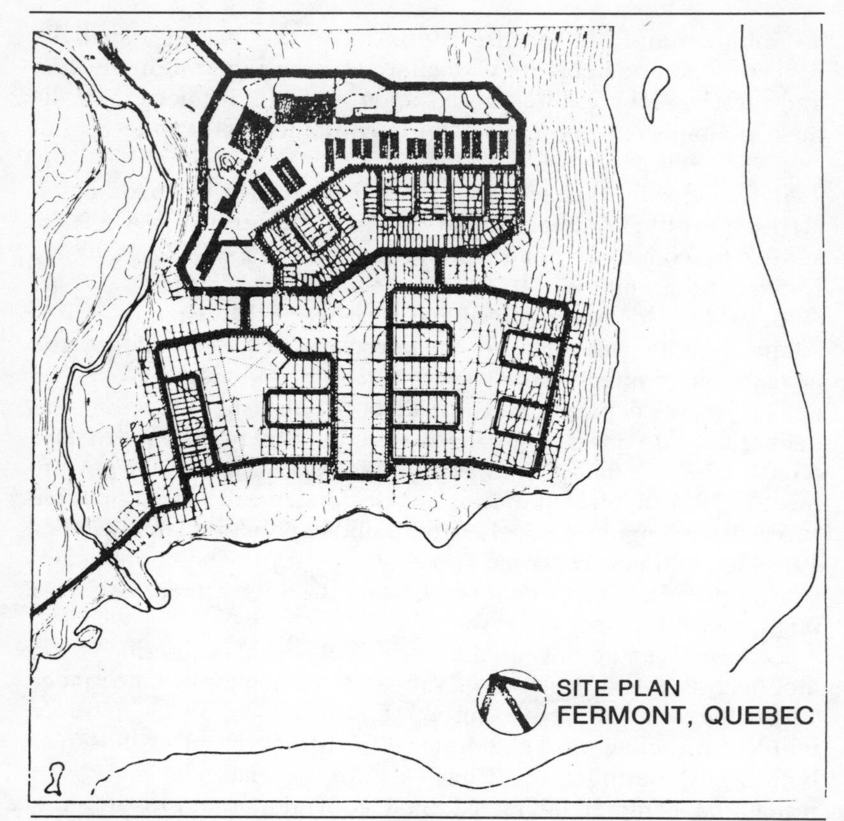

FIGURE I.1: Fermont, Quebec—a New Version of the Company Town

According to one climatologist (Phillips, 1985), if the "greenhouse effect" materializes in the next century, the consequences for Canada could be quite astounding and would include "larger Arctic cities, prairie orchards, shallower Great Lakes and submerged Atlantic and Pacific shorelines."

In the future a strategy for preparing for future climate changes may be an essential element of urban policy.

SIBERIA AND SOVIET CITIES

Siberia is slightly bigger than Canada both in land size and in population. As the more accessible sources of raw materials in the European part of the Soviet Union become exhausted, the Russians have been turning more to the east, across the Ural Mountains, all the way to their Pacific Coast.

Siberia already has eight cities with populations of over 500,000 people. The largest is Novosibirsk (1.4 m) in western Siberia. Novosibirsk, meaning "new" Siberia, has become both a manufacturing and intellectual center. Other Siberian cities are more related to the extraction of ore, natural gas, coal, and timber. Since 1984, when a second trans-Siberian railway was completed, the prospects for Siberia's development have been greatly enhanced, and its urban population will be increasing in the decades ahead.

One of the major planned Soviet cities—a "new town"—is Akademgordok, known as the Siberian city of science (Smith, 1978), with a population of 50,000. Designed with a large enclosed shopping center surrounded by apartment blocks and laboratories, this city represents an interesting Soviet effort to provide an optimal urban environment for some of its scientific elite.

Elsewhere, another writer (Bates, 1980) reports that "the cities are growing rapidly and agglomerations are now the dominant feature of the Soviet urban system." It is also indicated that in Soviet cities the customary central city functions are downgraded with the conscious decentralization of administrative and distributive services into fully developed secondary centers.

Although the patterns of Soviet urbanization are not very clear to most Western scholars, the dominance of winter in Soviet society requires at least this passing notice of their potential interest in the concept of livable winter cities.

In a recent book entitled *The Siberians* (1970), the controversial Canadian writer Farley Mowat has described life in primarily urban Siberia. Mowat is quite enthusiastic about life in the Arctic and has also written a book called *The Polar Passion* and a book in defense of wolves. He ends his Siberian book with these words:

> One thing is indisputable. Soviet physical accomplishments in Siberia are unmatched for their brilliance of conception and execution. The Soviets have, in not quite half a century, attained effective mastery over the entire region and are now able to direct its almost inconceivable potential towards their version of progress [Mowat, 1970: 358].

Another account of life in urban Siberia can be found in *I Saw Siberia* by the German writer Hugo Portisch (1972).

There is also a joint Arctic Development Commission between Canada and the Soviet Union, which recently has been revived after a suspension of activity following the invasion of Afghanistan in 1980. The development of the Arctic and Subarctic regions is beyond the scope of this particular volume but the planning for winter communities in those areas has become a very significant subdiscipline in Canada, Sweden, Norway, Finland, and the Soviet Union. For them the Alaskan motto "North to the Future" holds a special meaning (Rogge, 1973).

SECTION OUTLINE

In the first chapter in this section Pressman elaborates an extensive overview of the problems and prospects for winter cities. Drawing upon both his Canadian and European experiences, he develops a number of useful planning guidelines for the improvement of urban environments in cold geographic areas. Although many of his proposals and solutions involve architectural and urban design innovations or modifications, Pressman is also mindful of the social and cultural context of winter and is an advocate of the midwinter community festivals and vigorous development of alternative recreational activities in the wintertime.

In Chapter 3 Zepic presents five propositions concerning the nature of livability in northern climates before describing the development of Toronto. Although Toronto has the same latitude location as Rome, its climate is similar to that of Helsinki. She proposes a number of strategies to improve the microclimatic characteristics of winter cities through social, economic, physical, and administrative measures.

Metro Toronto is in fact a federation of six municipalities established in 1953 and Zepic also reviews their development strategy, which was prepared in 1981 and extends to 2001. Its concern for the evolution of a multicentered urban structure is an appropriate model for metropolitan growth elsewhere in the Snow Belt. The master plan for Toronto is currently under review and it is expected that the revised document will reflect new concerns about winter livability, including the extension of their underground mall system. Zepic also discusses the the impact of the Eaton Centre, which has become the model for enclosed downtown shopping centers in many U.S. centers.

Mänty discusses the development of cities in Finland. He notes that fully one-third of the world's population that lives north of the 60-degree latitude lives in Finland. The famous Tapiola Garden city is cited as an example of how the need to carry away snow was eliminated when buildings and streets were placed according to the topography of the area.

Both Sapporo (Japan) and Anchorage (Alaska) are discussed as winter cities that have experienced dramatic population growth in recent years. Masai reminds us that Japanese literature treats snow rather "warmly" and that for the average Japanese, snow has been something nostalgic, fancy, and elegant even though—or perhaps because—they live in areas where snow does not fall.

In Japan the Snow Country, or Yukiguni, is a beltlike region extending from western Honshu northeast up to Hokkaido where Sapporo is the cultural and economic center. In one sense Masai indicates that this northern island represents the "Western Frontier" of Japan, which ironically attracts migrants fleeing the summer heat of the rest of Japan. Sapporo is also seen as a city that incorporates more modern life patterns attractive to younger Japanese.

Sapporo hosted the 1972 Winter Olympics and continues every February to conduct an annual snow festival that is renowned for its international snow and ice sculptures. Besides extensive underground shopping plazas, it also maintains a sidewalk heating system. There is also an elevated transit line covered by semitransparent plastic for several miles. The subway line extends to ski facilities right within the city limits.

Anchorage's population tripled in the last decade and is now approaching 250,000 people, about half the population of Alaska. Johnson and his colleagues analyze the durability of Anchorage in relation to other cities as part of an analysis of quality of life criteria, and conclude that climate has little to do with growth.

The transactional nature of the urban economy of Anchorage is discussed, making this chapter a precursor to the section on transactional cities. Anchorage is more like a Canadian city than most U.S. cities in that it serves as a primary brokerage center for a land vastly abundant in natural resources. As the fastest growing city in America, the authors propose that Anchorage needs to better understand and utilize its transactional forces.

Another leading U.S. winter city, Minneapolis, is described in terms of its community context by its mayor and one of his deputies. Besides its skywalks, this city is known for its innovative political style, which dates back to Hubert Humphrey and the 1940s. Its community climate of mutuality is a good reflection of how people living in northern latitudes learn to pull together.

REFERENCES

BATES, J.H. (1980) The Soviet City: Ideal and Reality. Beverly Hills, CA: Sage.

COLLYMORE, P. (1982) The Architecture of Ralph Erskine. New York: Granada.

EDMONTON, B., M.A. GOLDBERG, and J. MERCER (1985) "Urban form in Canada and the United States: an examination of urban density gradients." Urban Studies 22: 209–217.

MERCER, J. and M.A. GOLDBERG (1974) "The fiscal conditions of American and Canadian cities." Urban Studies 21: 233–243.

———(1986) The Myth of the North American City. Vancouver: University of British Columbia Press.

MOWAT, F. (1970) The Siberians. Boston: Little, Brown.

PHILLIPS, D.W. (1985) "Canada's climate impacts program." Land 6, 1. Available from Lands Directorate, Environment Canada, Ottawa, K1A OE7.

PORTISCH, H. (1972) I Saw Siberia. London: Harrap.

PRESSMAN, N. (1985) Reshaping Winter Cities: Concepts, Strategies and Trends. Ontario: University of Waterloo Press.

RODGERS, W.C. and J.K. HANSON (1980) The Winter City Book: A Survival Guide for the Frostbelt. Edina, MN: Dorn.

ROGGE, J. (1973) Developing the Subarctic. University of Manitoba, Winnipeg.

SCHOENAUER, N. (1976) "Fermont—a new version of the company." Association of Collegiate Schools of Architecture Journal 29 (February).

SIMMONS, J.W. (1976) "The evolution of the Canadian urban system." The Usable Urban Past.

SMITH, D.M. (1978) "Siberian city of science." Geographical Magazine (August): 238.

Time (Canada) (1976) "The greening of Toronto." (June 23).

2

The Survival of Winter Cities: Problems and Prospects

NORMAN PRESSMAN

□CIVILIZATION COMMENCED AT the 70°F isotherm and gradually moved outward to more extreme climatic zones. Natural forces were conquered through the use of abundant energy supplies and available technological know-how, supported by adequate financial resources and institutional mechanisms of both the public and private sectors. Typically, the influence of climate has been largely ignored in project planning and policy development. All too commonly we have resorted to importing forms suitable for one climatic region to a completely different one. Although the influence of climate is not, nor should it be, deterministic, it should nevertheless act either as a facilitator or as an inhibitor of certain types of cultural adaptation expressed through built form. Climate must be regarded as a significant modifier of the built environment—but this has infrequently been the actual case in most of urban Canada or other Nordic countries. In fact,

> being a basic element of the natural environment, climate *is* one of the parameters of all architectural and urban design. The more extreme the climate is, the more necessary it becomes to respond to it [Culjat and Erskine, 1983].

It is most unfortunate that climate generally has been ignored in the cold regions, which typically also engender highly developed economies and technological prowess. These two features—money and technology—have been the major means whereby inhospitable environments can be either restrained or overcome. However, the vast majority of examples of climatic control occur at the building scale or micro level, for example, more effective forms of insulation, new roof panels and wall systems, double- and triple-glazed fenestration, and increasingly sophisticated hermetic seals. It is infrequent that the macro level urban pattern has been reshaped (that is, transportation systems, land use, and related elements). It is rare, too, that large-scale urban patterns are even examined from the perspective of climatic concerns and thereby viewed with the aim of creating a climate-responsive environment. As we are no longer living in an era of cheap and plentiful land and energy, we shall have to plan and manage our cities using a model of urban settlement that is highly integrated with the natural forces.

The "indoor-living period" in some northern latitudes has been estimated to be as high as 70% of total annual hours. Some studies have indicated that during the long winter months the majority of northern residents (sub-Arctic regions) spend as much as 95% of their time indoors. With so much time spent indoors, it is particularly important to maximize the positive aspects of contact with the outdoor environment by extending the outdoor season and by optimizing the beneficial climatic effects.

Socialization patterns in the urban north are different from those in the south. Hence, plans, designs, and policies should not anticipate equal intensity of provision or use of facilities for the population on a year-round basis. Some activities and forms of socialization even disappear during the very cold periods (January-April) when people tend to spend more time in the home than out and tend to engage in organized activities (work, cultural events, indoor sports, and even increased television watching). The rhythm of urban life often varies with respect to climate and the different seasons. The northern lifestyle is seasonally variable and is highly reflective of "climatic reality." This characteristic makes sense and should be retained to a limited extent. However, the possibility of increased outdoor contact can be created through proper landscape design and architectural/urban design strategies that acknowledge northern climatic place and strive to create better outdoor comfort for inhabitants.

Although it may not be advisable to overprotect urban dwellers from the cold because provision of too many "artificial" indoor spaces would prove economically unfeasible and perhaps even socially undesirable, it is nonetheless imperative to offer a modicum of protection from excessive negative stressors. A healthy exposure within the optimal range of comfort-stress scales will result in improved states of both physiological and mental health.

Although chance conditions will always be responsible for making some winter cities more livable than others (for example, location, microclimatic advantages), innovative, universally applicable planning and policy development measures can be instituted and are beginning to be evaluated. The approaches adopted in one community can often be transferred to another. The common component is the *adaptation of environments to extreme conditions of cold in the struggle to create living conditions that are better than merely tolerable.*

Regarding human adaptation to the cold, two fundamental approaches have evolved in northern latitude nations. They are as follows:

1. *Do Not Overprotect People from Nature.* This approach assumes that humans must learn to coexist with nature as satisfactorily as possible. If offered too much protection from the harsh elements, the human race living in cold climates will become too docile and sensitive instead of becoming adaptive, sturdy, and able to endure nature's inconveniences without heavy reliance on technology.
2. *Offer as Much Protection as Possible.* This proposition suggests that a wide range of human-made sheltering devices (tunnels, skywalks, atria, gallerias, etc.) should be incorporated within the existing urban structure so that minimal contact with undesirable weather systems is possible. It could be inferred that humans prefer "soft," protective environments as opposed to forced contact with harsh wind and temperature conditions.

These are two extreme opposite positions suggesting that *provision of choice* is critical to planning for northern climates. There is a unique beauty intrinsic to winter, but not all urban dwellers will be able to appreciate this beauty; for example, the elderly, handicapped persons, those with medical problems, people who are extrasensitive to the cold, and so on. One should therefore have the choice of being outdoors or withdrawing into warm, protected recesses either inside buildings themselves or in "urban pockets" that trap the sun's rays. It is this range of choices that must be analyzed as well as the ways in which they can be realized.

There are many complex faces to winter, especially in nations such as Finland or Canada where this season appears interminable. Harsh climate has worked its way into the national psyche of cold nations. Canadians have been depicted by others—and image themselves—to a very huge degree as a product of climate. The northern bleakness, with its cover of ice and snow, has repeatedly been a central theme in both French-Canadian and Anglo-Canadian poetry and literature. On the whole, our culture works hard in attempting to deny this hostile season, although at times we also delight in the snow-reflected light, the visual beauty, and the outdoor sports, carnivals, and festivities made possible when the landscape is snow covered. Although many of these carnivals—especially throughout Europe—have been rooted in religious rituals dating back to the Middle Ages, and today assume a more commercial and tourist nature, they have a distinct cultural derivation and meaning.

Winter and the problems it creates tend to be perceived in various ways. Solutions are found in many forms. For some, it is a matter of dressing properly—to fit the demands of the weather. We need better knowledge of the role of clothing—and even diet, for that matter—if we are to minimize the unpleasantness of extreme cold. What has been referred to as "weather-sense" must obviously be part and parcel of "cold culture" know-how. We must also be capable of systematically addressing the physical, social, and economic issues and be able to designate priorities and strategies for them at a range of hierarchical levels and scales of action. We must accept, respect, and work with the climate if cities in northern latitudes are to survive, and we must develop appropriate policies and urban design measures.

It is the intent of this chapter to address in a broad fashion some of the above-mentioned questions so that useful guidelines can be formulated for improving environments in cold geographic areas.

THE URGENT NEED FOR MAKING WINTER HABITATS MORE HUMANE AND CLIMATE RESPONSIVE

There are four basic reasons why more climate-responsive winter habitats are essential, within the Canadian context, but also elsewhere:

(1) Urban development and regionally oriented resource-based activities in cold climate areas can be expected to increase in the future as population pressures and the need for raw natural resources multiply.
(2) Traditional inner-city urban and social problems must be avoided. As well, future dilemmas must be anticipated and circumvented as the migration of populations from rural to urban settlements is an

ongoing process. Human needs and values must be reintroduced within the urban framework. Some of these include:

(a) making the city environment more humane by stressing more rationally planned urban development rather than urban sprawl, and

(b) turning the city into a community, the subdivision into a neighborhood, and the shelter into a home.

(3) The Frost Belt has been given a negative image (whereas the Sun Belt has been viewed more positively). A shift in this emphasis would encourage northerners to enjoy their places of permanent residence to a greater extent, to retire at home, and would encourage Sun Belt dwellers to experience winter vacations. In addition, this shift would make businesses and plants relocating to other areas realize that the Frost Belt can potentially be viewed as an attractive place for their employees.

(4) In southern Canada, where the majority of Canadians both live and work, winters can be relatively harsh and lengthy. In northern Canada, under even more stressful conditions, resource development has led and will continue to lead to the establishment of new communities—some of a permanent nature, others of a temporary one. The development of Canadian technological expertise in the field of severe climate adaptation would thus not only benefit both northern (and southern) Canadians, but would also provide a stimulus to our industrial and scientific enterprises in research, development, and production of goods having markets at home and abroad.

THE CULTURAL DIMENSIONS OF WINTER

Winter exhibits many faces in the world's different nations, not the least of which tends to be the animation of public life emanating from the carnivals or festivals. Although today such carnivals take on a distinctly commercial character, most were rooted in ancient religious rituals and traditions, of which a few will be cited herewith.

THE FASNACHT OF BASEL

Basel's most important and best-loved popular festival, the Fasnacht, commences each year at the stroke of 4 a.m. on the first Monday after Ash Wednesday, with the so-called Morgenstreich. At that time, all of the city's electric lights are extinguished while hundreds of lanterns are lit up, carried by masked and costumed merrymakers playing one tune in unison on high-pitched pipes and droning drums. As the night stretches into the dawn, the lanterns cease their glow, but the music and masquerading continues for three days as the inhabitants of Basel celebrate their winter carnival.

The traditions of the Fasnacht go back to ancient Babylonian rituals performed to drive out winter. During the Middle Ages, Christians incorporated the carnival into their calendars as a pre-Lenten festival.

Nearly every town is turned topsy-turvy at this time with fools becoming priests, men dressed as women, women as men, and youths enacting satirical skits at the expense of their elders. After the Reformation, most cities abandoned the carnival, though in some it evolved into a drunken revelry. Basel's Fasnacht with its mixture of costumes, otherworldly music, and political satire cleverly written into punning verses is one of the few remaining traditional carnivals. Neither relic nor riot, it provides Baselers the opportunity to express themselves and to ridicule, in public, controversial situations familiar to the citizenry. During the Fasnacht, no one dares to prosecute the "imitators" who are considered to be outside the law during the festival's duration.

In principle, the Fasnacht revelry is representative of a large number of colorful pagan rites and cults that are remnants of earlier cultures and epochs. Many of these cults were responses during the early Medieval period to the mysteries of birth, death, and the changing seasons, and mysteries deeply rooted in the minds of the people. In particular, a large number of pagan rites pertaining to fertility, ancestry, and vegetation were performed during the period of the dark nights between November 11 and February 2. These rites were eventually integrated within the church—in modified form—and a performance of pagan cults was allowed during the week prior to Ash Wednesday and was followed by Lent, with its 40 days of penance, fasting, and chastisement, even though the church attempted to eliminate them. Such are the origins of the Basel Fasnacht.

OTHER PRE-LENTEN CARNIVALS

In general, February is carnival time in most "Christian" societies. One of the famous events of all such Pre-Lenten affairs is the Carnival of Nice, which began as far back as Roman times. This important capital of the French Riviera shakes off its "mild winter" doldrums in the two weeks preceding Ash Wednesday, the traditional beginning of Lent. The samba dancers of Rio's streets and the high-jinks of the Mardi Gras in New Orleans hit the streets during the same period that the Trinidad Carnival is also celebrated.

Participants spend considerable time and money preparing costumes in all these places, and these periods are promoted as special events by the various Chambers of Commerce and Tourist Boards throughout the world.

THE CARNAVAL DE QUEBEC

One of the world's biggest *winter carnival* celebrations is "Le carnaval de Quebec" also held during early to mid-February in the ancient city of Quebec. This has become an event of international stature. Its recreational and cultural activities, sports events, parades, shows, nighttime festivities, and masked ball, presided over by none other than the renowned showman "Bonhomme Carnaval," make for 10 days of spectacular and exciting cheer. On the last day, the celebrated canoe races between Quebec City and Levis are conducted. Over the icy waters of the Saint-Laurent River the many participating teams compete in races and also revive a now-forgotten means of transportation that was an ancient tradition in Quebec.

All of Quebec is caught up in carnival fever. Huge ice palaces are sculpted (not unlike those found in Sapporo, Japan) within the framework of an international ice-sculpting competition. Dancing goes on in the streets and dogsled races are held. The festivities have been referred to as "Mardi Gras on Ice" and although some similar festivals were held in Quebec during the 1800s, the official carnival has been held annually since 1954. While all of the merrymaking continues, local inhabitants warm up during the city's extreme cold by drinking a local brew called *caribou*—a bracing mixture of red wine and hard liquor.

When the carnival is over, the city remains a delightful winter city, continuing the events initiated at carnival time. Citizens ice-skate on the frozen Charles River as music is played; they go tobogganing on the giant slide beside the Chateau Frontenac Hotel, and they snowshoe all over the city. Cross-country skiing is enjoyed in the Parc des Champs de Bataille at the Plains of Abraham and downhill skiing takes place in the Laurentian mountains, about a 40-minute drive from the city.

WINTERLUDE—BAL DE NEIGE (OTTAWA)

Judging by the number of winter festivals that have been springing up lately throughout Canada, there appears to be a lesson here for Canadians—that winter is worth celebrating. Under the aegis of the

National Capital Commission, the City of Ottawa has in recent years been sponsoring its version, called "Winterlude," held at the same time as other winter carnivals. People skate on the Rideau Canal, site of the world's longest skating rink—roughly five miles in length. In addition, races and barrel jumping competitions are held while snow sculpting and sleigh rides round out the merrymaking. A grand parade, "February's March," animates the streets with floats, bands, and red-nosed clowns who wind their way through the illuminated corridor of the canal pausing to watch the frozen fireworks in the sky. *Piruvik* (which means "growing place") is a winter playground designed especially for children during this period. One of the most unusual horse racing events in the world, the "Canadian Club Classic," is run on the Rideau Canal. Additional events such as the Canadian Ski Marathon are held—a two-day, 160-km cross-country endurance race from Lachute to the Canal.

Other festivals exist throughout Canada, such as the Northern Manitoba Trapper's Festival at The Pas, which celebrates the good, hard life of the north with vigorous outdoor competitions including dogsled races, music, food, and drink, and a revival of games and sporting events derivative of the fur-trade era. Carnivals seem to be "warming up" the winter and they are beginning to attract tourists from more southerly regions, especially the United States. Furthermore, they not only assist in relieving the "February blahs," or winter doldrums, but are coming to be viewed as good business for communities throughout the nation.

PROTECTING ONE'S BODY FROM THE COLD

Frequently, we have not understood how our bodies work when confronted with the effects of cold weather. Wearing the wrong clothing or drinking the wrong fluids may at worst be risky or at best lead to severe discomfort. The human body—a "heat machine"—must be maintained at a fairly even temperature (37° C) regardless of the ambient temperature surrounding it. Should the air be colder than its internal temperature, the body will lose heat and must invoke defenses against the cold by taking in more food, by being better clothed, or through engaging in muscular activity that may include work, exercise, or even shivering. Shivering is one way the body adapts to cold, by making us move our muscles even when we're standing still. It is an involuntary reaction, as nobody decides deliberately to shiver at a given point of time. Shivering is an automatic response to a situation in which the body's internal temperature is getting too low.

There are dangers even when the body is working at its optimum. One such danger is frostbite—usually directly affecting parts exposed, such as ears, noses, or the extremities (feet and hands). Because the feet are normally exposed to colder conditions than the rest of the human body, they are prime candidates for frostbite. Hence, during intensely cold weather all exposed parts of the body should be examined to see whether they appear stiff or without feeling. If numb and accessible, they should be immediately warmed, covered with the hands, or held close to the body of a companion.

Good habits help to keep one warm. A poor diet, nicotine, alcohol, and lack of adequate sleep will undermine the body's own natural ability to adjust to cold weather conditions. In extreme cold, the body requires a longer warm-up period—not unlike that of an automobile. Prior to going outside or getting into ski gear one should indulge in about 15 minutes of jumping-jacks, push-ups, or running in place. This will minimize the shock of exposure to the outdoors.

Wearing the correct clothing can protect us from the many unpleasant effects of cold weather. For example, layered clothing is much better than one thick garment as the layers tend to trap air, which in itself is a good insulator. Closely woven wool garments are good insulators, as are heavy wool socks (cotton socks worn against the skin, under the wool ones, act to better insulate one's feet from the cold even when they are wet with perspiration).

Minimizing the windchill factor is essential as this phenomenon can be very dangerous if one remains outdoors too long or is improperly dressed. A strong wind can make even a mild day feel bitter as wind blows away the warm air lying beneath clothing or between its layers. For example, a gentle breeze of not quite 2.25 miles per hour (3.62 km/hour) carries off three times the heat from one's body as would be lost if the air were stationary. Windchill can make the weather feel much more unpleasant than the temperature alone might indicate. In fact, windchill is one of the major causes of heart attacks and frostbite in cold weather.

Weather has a powerful influence on how we feel, how we behave, and how our bodies perform to counteract nature's impacts. It is imperative to understand the effects of temperature, windchill, humidity, and other related factors. Fewer weather-related mishaps would occur if this "common weather sense" were part and parcel of the know-how of dwellers in northern latitudes. It should become second nature to the inhabitants of Snow Belt regions and must be addressed through public education programs if our survival mechanisms are to be more effective in combating the cold.

PERCEPTIONS OF WINTER PROBLEMS

In an attempt to identify more explicitly winter-related problem areas, I undertook a survey (in collaboration with a Livable Winter City Association colleague, Xenia Zepic, of the Metropolitan Toronto Planning Department, Policy Development Division). The project, which was initiated during the fall of 1984, was called "Developing Livable Winter City Environments." Questionnaires were sent to planning directors throughout Canada in municipalities having a minimum of 50,000 inhabitants.

Some of the more salient issues requiring urgent attention, as indicated by the survey's results, were indicated to be the following:

- lack of attractive outdoor amenity areas;
- lack of enclosed public open spaces (providing shelter and heat);
- lack of enclosed parking areas in the central business district;
- inadequate snow-removal service (for roadways and sidewalks);
- need to minimize windchill and snowdrifting;
- need for guidelines and standards acknowledging winter climate policies;
- improved public transit facilities, including well-designed bus shelters;
- enhanced mobility of the disabled;
- financial assistance grants for heat and light to low-income groups;
- reduced energy consumption and dependency;
- greater attention to groups that are more vulnerable to extreme climatic conditions, such as poor, elderly, handicapped, and so on;
- contingency measures to accommodate emergency conditions, for example, severe blizzards, heat and power failures, and so on;
- encourage citizens to appreciate and utilize their natural winter environment, for example, stimulate outdoor participation—snowshoeing, cross-country skiing, and so forth;
- urban pattern and architectural form to reflect particular climate zones;
- residential areas to provide covered zones, for example, roofed-over links or interconnected buildings;
- greater analysis of wind-induced discomfort through wind-impact statements;
- encourage district heating systems where feasible;

- reduction of travel demand and required distances (by transit, car, and foot);
- encourage innovation through existing mechanisms and assistance programs;
- attract greater attention to "winter dilemmas" by the public at large;
- diversify economic base where abnormally high seasonal unemployment results during winter, such as those found in resource-based towns;
- revise building codes, bylaws, policies, and development guidelines so as to reflect "winter conditions" where this is not already being done;
- support in-depth research and experimentation;
- improve outdoor street illumination systems with variable control for extreme weather conditions;
- enhance aesthetic quality by the use of more color in winter; and
- improve the design of winter clothing and gain greater public awareness of how to dress for extreme cold conditions.

Although these responses are not to be interpreted in any order of priority, they are nevertheless reflective of the general disregard among various levels of government for the multiple problems generated by the winter season. They tend on the whole to focus on physically related matters, but these have strong economic and social implications. They address a range of scales—from the small detail such as design of more user-responsive bus shelters, to broader components such as land-use and transportation relationships, energy-efficient layouts and policies, as well as the need for further research and analysis that must identify hierarchies and priorities of respective levels of action and intervention.

As Hans Blumenfeld (1985: 50) has so eloquently stated:

> God lives in the detail. So does the devil. We may never succeed in driving him out of all the nooks and crannies and create The Livable Winter City, but we certainly can do better than we do now to make our cities more livable in winter.

IDENTIFYING A CRITICAL RANGE OF WINTER-INDUCED DILEMMAS

In the majority of instances, our modern northern urban environments turn the winter experience into anything from nuisance to nightmare. The low angle of winter sun has great difficulty penetrating buildings and open spaces, with shadows often cast 15 times the height of the buildings. Snow melts into slush creating driving hazards and severely inconveniencing pedestrian movement. Windchill makes one feel much colder than the air temperature alone might suggest. Lengthy exposure to the cold—for example, waiting 20 minutes for a bus in −20°C, which can produce frostbite—is debilitating and extremely unpleasant, if not painful. Walking on icy sidewalks can be dangerous, especially for the elderly and handicapped. Shoveling snow in order to get out the front door or take the car out of the driveway is known to have increased the rate of heart attacks. Accident rates climb steeply in urban and rural areas on roads and expressways during heavy snowfalls and blizzardlike conditions.[1] The number of socially oriented activities is greatly reduced during the cold season. It has been established that even on "nice" winter days of relatively mild temperatures and low windchill factors, the peak numbers of people outdoors are far below the summer averages (Nash, 1983: 29–30).

Spontaneous activities such as play, walking, or simply "hanging out" are the first to be affected by the onset of winter conditions. Public outdoor spaces that are heavily utilized during the late spring and summer are transformed into scarcely used circulation paths in winter.

Despite these inconveniences to human well-being and the concomitant social risks, architects, planners, and policy analysts still insist on orienting their designs and development policies toward the brief summer season, especially with regard to public outdoor space and parks. A great deal can and must be done to moderate the severity of microclimate or to prolong the season of comfortable outdoor activity through the effective use of techniques such as windscreens, planting, and southern orientation. Buildings can act as barriers protecting one from the wind, "wind gates" can be set up to reduce the wind velocity on downtown streets, and proper planting (positioning of trees and bushes in appropriate locations) can serve as "wind sponges" to reduce windchill. In many instances, it is simply a case of using nature to protect one from nature.

Foremost, it is essential to identify a range of critical issues requiring high-priority attention that must be addressed by both the public and private sectors. These categories shall be defined as follows:

PHYSICAL

Pedestrian protection—colonnades, arcades, through-block linkages, galleries, atria, and integrated bus shelters.

Access optimization—reduction of walking distances to transit facilities, parking areas, and major retail and institutional complexes.

Integrated development—guidelines and policies (offering incentives where possible) promoting improved urban designs, higher density mixed uses, energy efficient urban patterns, utilization of microclimatic principles, intensification of residential neighborhoods, and transportation corridors.

Public open space—seasonal approach to the design and use of civic spaces and neighborhood park systems so that greater use occurs throughout each season. Multifunctional use is the key issue. The creation of winter gardens and indoor parks at strategic locations should be encouraged in the form of pilot projects.

SOCIAL

Safety and health care—programs and public education with a focus on winter clothing, survival techniques under emergency conditions, winter driving skills (skid-control schools), and winter fitness programs.

Winter promotion events—projecting positive images of winter, festivals, carnivals, ice sculpture competitions, winter sport displays, and winter innovation contests.

Community snow clearing—collective activities related to snow clearance (and related social encounters) and assistance by volunteers to senior citizens and the handicapped.

Policies for special needs groups—more efficient location criteria for senior citizens homes, at-home delivery of convenience goods, and special mobile vans to get nonmotorized inhabitants out of the house for medical reasons, social events, or other tasks.

ECONOMIC

Transit assistance—fare reductions on certain "winter sale days" under the auspices of a merchant's association or "business improvement area," and special vehicles (4-wheel drive) available when transit is severely impaired due to snow accumulation or icy road surfaces.

Winter subsidy programs—special grants or tax deductions to low-/moderate-income households for heating, lighting, winter clothing, and upgrading of insulation in owner-occupied homes.

Tourism promotion—grants and incentives to induce winter-oriented recreation and tourism, carnivals and festivities, winter landmarks, and historic sites.

Employment and retraining programs—especially crucial when unemployment increases due to seasonal variations in economic activity, as in the construction industry. Resources needs to be redirected toward public works tasks, street maintenance of higher quality, and recreational facilities management in both urban and nonurban areas.

AN ATTEMPT AT RECTIFYING WINTER'S PROBLEMS: WHAT CAN BE DONE?

What can be done about "winter malaise," "February blahs," or "winter hibernation"? Where should we commence if we are to minimize winter-induced stress and inconvenience in the daily urban routine? If we are to exact a truly pragmatic posture in answering these questions and acknowledge the fact that we are living within an economic framework that tends to be recessionary in nature—with public and private sector budgets extremely tight or gradually shrinking—then we must look for answers in the detailed improvements that can be implemented on a continuing, even piecemeal basis, but that must form part of a broader, more comprehensive strategy. These can usually be achieved at the local level (neighborhood, town, city) through combined efforts of individual citizen groups and organizations, local business associations, and municipal government. Higher-level governments need not be relied upon, except where funding assistance and special granting programs are available for joint action projects.

A few examples should indicate a general direction of what can occur.

(1) *Bus Shelter Design*. Comfort, convenience, and accessibility must be significantly increased vis-a-vis public transit operations. More bus shelters must be provided (there are still many cities in cold climates that have no bus shelters at all, even in the central shopping precincts). These must be at strategic locations throughout the central area, neighborhood shopping developments, and residential zones. They should be *heated* (by solar or radiant means). They should provide ease of maintenance and access, adequate seating, comfortable standing room, bus routings and schedules, public information, trash receptacles, outdoor benches (for nice days), and high-quality illumination (especially at night). We sorely require much more than

the simplistic "glass booths" occasionally found in many large cities, but hardly in small- or medium-sized towns.

(2) *Pavement Heating.* Stairways and pedestrian ramps could be electrically heated where benefits would accrue to users—especially the elderly or handicapped. Highways and roads could also be heated at difficult intersections or where steep grades are present, so that snow and ice would melt and hazardous driving (and walking) conditions would be alleviated. Considering all costs—direct as well as indirect—of municipal snow clearance plus salting and sanding operations, such an option might be worthwhile of serious consideration. This has been done in Sweden (especially in Stockholm's business district) where the initial capital cost tends to be justified by future savings on both labor and snow removal. Overhead radiant heating could also be provided over intensively used sidewalks in heavily frequented shopping and entertainment zones.

(3) *Eliminate Street Curbs.* Street curbs at intersections could easily be eliminated. This is presently the case in many Swedish towns, both large and small. Town engineers have evened out the grade at street corners for pedestrians by raising it for motorists (thus acting as a gradual "speed bump" for automobiles). The pedestrian now benefits from the "least resistance" movement, with cars being demoted to second-class status. Eliminating curbs is appropriate to winter cities, for puddles accumulate at these locations and people can avoid stepping into water. They also will not get splashed by passing vehicles. Storm sewer inlets might have to be redesigned and relocated, but this would constitute the only impediment. It is an extremely simple device and would achieve much for pedestrians.

(4) *Trap The Sun.* Encourage more mixed-use activities and incorporate the climate-protection principle in designs and policies, trapping the sun wherever possible. Orientation of open spaces, skating rinks, park benches, and buildings should achieve maximum passive solar gain. Wind-impact statements should dictate site planning development to minimize wind tunneling (thereby reducing wind-chill factors) as well as indicate the nature and location of planting to absorb wind (acting as a sponge) and reduce its cooling effect. Experiments should be undertaken with respect to the creation of "winter gardens" and "indoor parks" for a variety of urban activities and different user groups with varying demands and preferences.

(5) *Develop Positive Attitudes and Think Winter.* Attitudes and policies to assuage the worst of winter are fundamental necessities and we are really being shortsighted if we do not acknowledge this obvious fact

(Scotton, 1984). We must "think winter" when formulating policies or conceiving plans and urban developments (however, designers usually "think summer" in their mind's eye over the drafting board, indicating open spaces with green hues when seven or eight months of snow and ice may cover the project site). We must develop what Ralph Erskine termed a "grammar for high latitudes" (Collymore, 1982) whereby architectural forms and development policies respond closely to the climatic forces at hand. We should no longer function exclusively with the standard urban development vocabulary of the southern United States or of Mediterranean Europe (enclosed urban squares or piazzas that tend to be unsuccessful commercially as well as socially).

(6) *Use Bold Colors.* It has been known for quite some time that the strategic use of color can achieve certain mood reversal effects. For example, the use of "cool" hues in hospitals or reformatory cells has been standard practice in many parts of the world. The color consultant Faber Birren has noted that

> yellow, orange and pink direct the attention outwards and create an environment conducive to muscular effort, action and cheerful spirit. . . . The cooler colours, that is, the blue end of the spectrum, minimize distractions from the environment and aid introspection, visual and mental tasks (good for offices, studies, babies' rooms, etc.)[Zrudlo, 1972].

Under severe winter conditions, the judicious choice of colors for a community can be helpful in identifying buildings, landmarks, or entire settlements—at great distances—during conditions of whiteout (caused by blowing snow), high solar reflectivity, or under intense fog cover. A range of colors might be selected for various building types in order to provide contrast under different seasons or climatic conditions. The brighter hues—red, orange, or yellow—are most easily recognized in daylight and present the highest contrast with snow cover. Because the winter landscape lacks color contrast a "balancing effect" must be introduced with emphasis on warmer hues in order to counteract the monotonous—and cold—effect of whiteness. The skillful planting of coniferous vegetation can be another way in which color can be introduced into an otherwise unstimulating landscape.

Color has been used most effectively on houses in Greenland. Jorgensen, describing this phenomenon, states that

> they are painted in different colours such as dark red, yellow, blue, brown and green, giving the town a pleasing appearance, as in the summer they are in contrast to a background of the brown-green grass and in winter they are in contrast to the snow [Zrudlo, 1972].

Color should be used in more bold and imaginative ways to cheer up our lifestyles, buildings, and natural surroundings. Northern towns should employ more color to minimize the drabness of winter. An extremely well-known landscape architect practicing in Toronto put it this way:

> Whether snow-covered or just plain grey and dreary, buildings are often lifeless and the use of colour in architecture has only been timidly pursued. In my own work I have been using the primary colours of red, blue and yellow to help articulate space and create focal points. The use of colours on fences, for example, can make drab garden spaces magical in winter. . . . Cadmium yellow on a west facing fence is particularly beautiful as the late afternoon sun literally "heats" the space and gives it a warm glow [Kehm, 1985].

Exemplary settings show

> the use of pennants on the exterior of the Guthrie Theatre in Minneapolis, the brightly colored houses of Quebec and Greenland, the Saxon gold, brick red, and Cape Cod green used on homes by our Colonial ancestors, and the gold domed churches of Russia [Rogers, cited in Ray, 1983].

The once popular idea—which evolved during the 1960s when Buckminster Fuller developed his geodesic dome experiments exemplified by the United States pavilion at the Montreal site of the 1967 International Exposition—of doming over entire cities or neighborhoods in order to master climate has given way to a broad spectrum of smaller, detailed, and more attractive solutions. They may not be terribly dramatic, but taken together they can contribute significantly to the way quality of life is perceived in winter cities. They are much less costly than "mega-schemes," can be implemented either quickly or over a long period of time, and can be extremely cost effective and cost sensitive.

POLICY-LEVEL RESPONSES OF A BROADER SCOPE

In addition to the more detailed, incremental measures that can be adopted to assist in alleviating winter's difficulties, there are numerous policy responses that can have a major impact when addressing the meso-scale (urban block, street, neighborhood, precinct) and macro-scale (district, town, metropolitan area). On the whole, such orientation deals with "winter management" and development strategies that, in concert, tend to produce energy-efficient urban structures with concomitant administrative responses.

Human behavior characteristics under cold conditions surely have powerful implications with respect to the behavior of bureaucratic structures, law enforcement agencies, public behavior that may or may not be tolerated, and similarly related institutional norms. *Winter utilization of cities might benefit from different approaches to policymaking than summer usage of the same cities and their facilities.* Time management ought to be given serious consideration with respect to how a city is used at various seasons of the year. Perhaps the reduction of headways between buses should be arranged during the harsh winter periods, for surely it is more debilitating to wait 20–30 minutes for a bus under the extremely cold conditions of the December through February months than it is during the summer. It is possible that the concept of the "electronic cottage," whereby more people will have the opportunities and choices of working in the home, will make winter habitation more bearable as the need to leave the home for a workplace—spatially separated from the dwelling—can be drastically reduced. Whether people will choose to accept such an option is debatable (Naisloitt, 1984), because people wish to be with people, but this might make much more sense in the Snow Belt regions than in the Sun Belt. How urban management techniques can respond more precisely to seasonal demands has not yet been made entirely clear and is an excellent subject for further study.

There are, nevertheless, several matters about which we can be relatively certain. These are areas in which urgent action is required. We need:

- more compact and energy-efficient urban development to be achieved through a reconfiguration of land use and transportation patterns;
- greater mixed land uses throughout the metropolitan region;

- more emphasis on transit that is both accessible and affordable;
- intensification of residential neighborhoods and a greater injection of residential fabric into the central business districts;
- sensitive redevelopment, based on climate-responsive principles, of obsolete industries occupying strategic industrial lands;
- additional transit corridors linking suburban centers to one another and to the central business districts;
- integrated developments whereby housing, commercial, retail, and transit facilities are connected by systems that protect users from inclement weather;
- development of suburbs, creating true urban nodes, housing and offices superimposed on or adjacent to shopping centers and connected to the central city not only by expressways but also by effective transit lines (Yorkdale Shopping Centre in Metropolitan Toronto serves as an excellent prototype and one of the few examples of its kind in North America);
- more comparative research of an international scope dealing with winter management policies; and
- more community education programs assisting people to understand winter and cope with it more satisfactorily.

Such is a conceivable "urban affairs menu" for future consideration. It is imperative to "think winter" when formulating policies, concepts, and strategies that hope to solve the problems of winter cities.

CONCLUSIONS

The aim of this chapter was not to provide "ultimate" solutions for specific communities or countries. Instead, it has been to describe an overview approach toward improving livability in cities from which various localities with their respective site and culture-specific issues can derive guidelines for developments and policies. It is such a conceptual basis that can serve as a framework for enhancing the quality of community life. If "cities in the cold" are to be reshaped in more responsive ways, three levels of consideration must be addressed:

(1) the micro-level (house, dwelling cluster, building detail),
(2) the meso-level (city block, street, neighborhood, precinct), and
(3) the macro-level (district, town, metropolitan region).

If strategies can simultaneously be implemented at all three levels of concern, we can expect a significant improvement in the way we experience and perceive our built environment. Some of these principles refer to elements such as favorable orientation of structures, wind protection, introduction of climate-modifying plant life, clustering of dwellings, mixed land use, urban infill, more rational distribution of jobs and places of residence, provision of more energy-efficient public transportation systems, and high-quality sheltering techniques such as covered sidewalks, heated bus shelters, enclosed streets, and underground development. Such concepts must be tested and evaluated with regard to cost-benefit considerations on fairly elaborate scales where this has not previously been done.

It is hypothesized that design and policy decisions should be related where possible to the behavioral sciences, if this can provide a more informed view. Nourishment received from this source can make a noteworthy impact on the improvement of urban living. Through interdisciplinary collaboration, designers and planners have the possibility of becoming more socially responsible. Reciprocally, behavioral scientists and environmental sociologists will have greater opportunity in the realm of applying their research findings in more practical ways.

With such joint activity, those working on the gap between social needs and environmental design responses—dealing with the interaction of humans, society, and the built environment—can more meaningfully contribute to development of the knowledge base affecting these two concerns. *Higher standards* of both *urban planning and policy development* can be achieved and, as a result, more effective social research and planning can be undertaken.

Updating, extending, and expanding the range of issues for which accurate predictions of use and behavior can be made is a constant research priority.

We have few broadly accepted norms for many routine issues—such as the amount of open space necessary around schools or housing developments, the amount of sunlight desirable at street level in built-up areas, and so on—despite the fact that regulations and policies commonly prescribe such matters.

We must continually improve standards by proposing new or revised norms based on studies, being explicit about where and why such norms would seem to apply. Furthermore, there is an urgent need to add new standards (or revise old ones) that reflect microclimatic and solar access concerns in site design.

How are the necessary improvements to be achieved? Apparently, one must work within the existing legislative frameworks of the municipalities and provinces, or central governments, if we are to anticipate realistic solutions. The two major techniques that the majority of cities utilize for implementing urban design and policy measures are the following:

> the development regulations that apply to new buildings and a strategic use of the city's own government funds and the outside grant money it is able to attract [Barnett, 1984].

These two methods are the determining factors in managing urban growth and change. Zoning bylaws, development controls, and official plan policies embodied in performance standards and development review criteria will set the tone for what may or may not occur, although it must be understood that they should be neither too specific nor too general if they are to achieve their intended effect, remain realistic, and still respect the performance characteristics and economic viability of the development proposal, without imposing unnecessary restrictions. Furthermore, "bonus densities" could be provided whereby developers would be encouraged to provide "winter amenities" and public-oriented equipment such as cross-country ski paths (in residential subdivisions), indoor parks and open space, pedestrian skywalks, underground tunnels, or mid-block arcades—in exchange for an increase in building coverage or other features attractive from the development perspective.

Future research must assist in resolving the debate on the nature of "winter problems" and "winter induced stress" so crucial in the northern planning context. In addition to answering questions of a more theoretical nature regarding the effect of climate on behavior, research should also contribute to the development of social and physical planning guidelines and interventions dealing with urban policy.

It must be assumed that problems can be resolved if both government and the private sector address them candidly and firmly. The range of approaches should be pragmatic with most being relatively undemanding of institutional adjustments or significant shifts in behavioral patterns.

We need to encourage more innovative, creative, and dramatic experiments and policies because

> in dealing with the north, what we have been doing almost unquestioningly is leap frogging southern development farther north. . . . If you go through the north, you will see almost no innovative solutions [Ministry of Municipal Affairs and Housing, Province of Ontario, 1981].

Finally, it must be emphasized that winter should not be seen as an isolated part of the year but as part of the total cycle. The amelioration of harsh winter conditions will not constitute improvement unless the amenities and shortcomings of all the seasons are considered in relationship to one another. We must maximize the theatre of seasons and the human-made environment in order to create comfortable, desirable, and efficient year-round cities.

NOTE

1. The resulting increase in the daily accident rate during snowfall conditions in metropolitan Toronto is in the order of 1.3 to 2.4 times the average daily rate. The proportion of accidents on collector and local roads is approximately 50% higher on snow or ice as compared to dry or wet pavement (Mende, 1982).

REFERENCES

BARNETT, J. (1984) "Urban design as a survival tool." Urban Design International 5, 1: 27.

BLUMENFELD, H. (1985) "Problems of winter in the city," p. 50 in N. Pressman (ed.) Reshaping Winter Cities. Ontario: University of Waterloo Press.

COLLYMORE, P. (1982) The Architecture of Ralph Erskine. London: Granada.

CULJAT, B. and R. ERSKINE (1983) "Climate-responsive social space: a Scandinavian perspective." Environments 15, 2 (University of Waterloo, Ontario).

KEHM, W. H. (1985) "The landscape of the livable winter city," p. 59 in N. Pressman (ed.) Reshaping Winter Cities. Ontario: University of Waterloo Press.

MENDE, J. (1982) "An analysis of snowstorm-related accidents in metropolitan Toronto." Master's thesis, University of Toronto.

Ministry of Municipal Affairs and Housing, Province of Ontario (1981) Building Toward 2001. Proceedings, pp. 29–30.

NAISBITT, J. (1984) Megatrends. New York: Warner.

NASH, J. E. (1983) "Human relations in frozen places." Architecture Minnesota, (January/Februay): 29–30.

RAY, J. (1983) "Colors of the northern winter landscape." Minnesota Horticulturist 11, 9: 272.

SCOTTON, L. (1984) "Our cities should be built for winter." Toronto Star (February 25).

ZRUDLO, L. R. (1972) Psychological Problems and Environment Design in the North. Collection Nordicana, Number 34, Centres d'Etudes Nordiques, University Laval, Quebec.

3

Toronto: Policies and Strategies for the Livable Winter City

XENIA ZEPIC

□IN 1976 Time magazine proclaimed Toronto the most livable city in North America. The recognition Toronto received was justified. This single issue (June 22, 1976) was responsible for putting Toronto on the world map and evoking a sense of pride among Torontonians for their city. It was the time when a newly elected radical council and its mayor, David Crombie (presently the Minister for Indian and Northern Affairs in the federal government), changed the course of development not only within the city of Toronto but also within the metropolitan Toronto area.

At that time the city of Toronto was the only major city in North America that had halted development in order to reassess the validity of its existing planning policies and reformulate new ones to improve the overall environment, emphasize preservation, and strengthen existing neighborhoods. It was evident that assumptions and planning principles that had governed development in the fifties and sixties were no longer acceptable in the mid-seventies. New plans and new directions that would reflect different attitudes and a different economic reality were necessary.

Changes in area municipal planning policies were reflected in the new Metropolitan Official Plan, which was formulated in 1980 (Metropolitan Toronto Planning Department for Metropolitan Council, 1983). Principles were adopted for restructuring land use/transportation relationships, energy conservation, and environmental protection in order to create a livable urban environment.

As the Provincial Planning Act encourages a regular review of area municipal plans every five years, metropolitan Toronto is presently embarking on a review of its Official Plan by testing assumptions regarding employment, population, and transportation projections and demands. Trends indicate that our cities, as much as our lifestyle, will change within the next two decades. Economic growth is expected to slow down, the population will become more aged, and high technology will replace most of the old industrial base.

Until now, climate and its influence on lifestyle, urban environmental quality, as well as on the management of cities, has received very little attention from planning or political bodies. With the disadvantages of living in a cold climate, and with technological means available to improve our living conditions, it is hoped that a revised Metro Official Plan will better articulate these concerns.

Within this context, only those policies in the Metropolitan Official Plan that are related directly to the definition of livability in northern climates are discussed. In addition, case studies of climate responsive projects that were built in spite of the lack of articulated legislative measures will be described. Jointly they illustrate the commitment on the part of the public and the business community toward creation of a truly livable winter city.

WHAT CONSTITUTES LIVABILITY IN THE NORTHERN CLIMATES?

The often used word "livability" has many meanings to different people. When used in the context of an urban area, it usually means an individual's satisfaction with the level of services and amenities. As each of us has different aspirations and make different demands on our urban environment, the perception of livability also depends on our own personal level of satisfaction.

In this regard, it is safe to say that livability is dependent on two interrelated sets of variables—the economic, physical, and mental well-being of an individual on the one hand, and the level of efficiency, safety, and beauty of the built and natural environment on the other.

In an economic sense, livability could be defined by job availability, the cost of living, the price of housing, and the ease of mobility. The physical and mental well-being of an individual is dependent on the availability of immediate medical care, plus the quality and proximity of educational, recreational, and social amenities.

This definition of livability is applicable to all built environments regardless of geographical or climatic conditions. However, extremely cold and hot climates impose restrictions on the behavior of the individual as well as on the form and economic viability of urban areas. Thus, it is within the context of the northern climate that the livability of urban settlements will be analyzed.

Proposition 1. Job opportunities and responsible management of economic resources should be a priority. The cities of North America are either the oldest or the newest depending on geographical location. Whereas recent "new-town" activities for Canadians have been concentrated in the far north, new settlements in the United States are located within the Sun Belt. The oldest ones are generally located within the Frost Belt of the United States, or along the Great Lakes in Canada. They are the result of industrial growth at the turn of the century, with infrastructure and housing stock slowly becoming obsolete in view of age and changing technology. Most of these industrial cities experience problems of seasonal unemployment during the winter months, in addition to loss of jobs due to the continuous shift toward information-based technology. It is estimated that between 40,000 and 50,000 workers are unemployed during the winter season annually between November and June in the metropolitan Toronto area. The jobs most affected are in construction, tourism, recreation, and some aspects of transportation.

The new resource towns in the Canadian north could be classified as single-enterprise communities. Development of those towns is based on nonrenewable resource exploitation and is characterized by successive booms and busts. Their existence and prosperity depends upon outside market forces, which in turn regulate production input and employment levels. Because of their single specialized economy, they easily become victims of changing world market conditions over which they have no control. The legacy of some of these boom and bust communities is that of desolate, uprooted people, and doomed settlements (Gibson, 1976).

According to Jane Jacobs, markets, jobs, technology, and capital transforms the cities' own regions. Their success or failure depend on their ability to adopt to new economic forces (Jacobs, 1984). Marketing strategies and development policies are tools that guide their economic growth. Their climatic characteristics should be a part of the process of determining the type of economic development as well as their physical and spatial distribution and urban form.

Proposition 2. More compact and diversified urban form would improve mobility, provide for better social interaction, reduce heat and transporta-

tion energy, and lower construction and maintenance costs. Although the amount of money spent on food and clothing is to a large extent dependent on the lifestyle of an individual, the cost of energy, housing, transportation, and recreation is related directly to the design and efficiency of the urban settlement. Housing and transportation are two sectors where considerable energy and cost savings could be gained. A survey of energy consumption in the average Canadian city shows that residential uses account for 35% of total energy consumption (McLellan and Underwood, 1983). On average, buildings throughout northern belt regions have to be heated at least seven months of the year, and the single-story detached house—the most common and popular housing form—is one of the biggest energy consumers. Although the cost of energy has to some extent already influenced changes in the lifestyle of many Canadians, especially among low-income families and senior citizens, it is not perceived as a significant enough factor to affect the prevailing forms of shelter. Suburbia, with its low densities and single detached homes, is well and expanding!

The private automobile accounts for two-thirds of the 25% of energy allocated to the transportation sector, whereas only 2% of energy consumption is allocated to public transit (McLellan and Underwood, 1983). In addition to large expenditures for fuel and gas, the individual is also faced with higher car maintenance costs due to road salting and more frequent accidents during winter months.

The costs of construction, management, and maintenance of infrastructure are also much higher in northern settlements. As an example, in 1984 metropolitan Toronto allocated $10.5 million for snow clearance and removal alone (Municipality of Metropolitan Toronto, 1986). The total network of arterial roads under metropolitan jurisdiction amounted to 764 kilometers or 3,133 lane kilometers, which meant that $13,750 per kilometer was spent on snow removal annually. Even a slight reduction in the length or width of roadways would significantly reduce the cost of winter maintenance.

For many people, the distance between their place of residence and their workplace and the resultant long commuting time is increasingly becoming an issue. A mix of various activities and employment opportunities in close proximity to residential areas would provide greater choices available to persons who wish to reduce their commuting time. This pattern would also contribute to the reduction of municipal expenditures related to the physical form and upkeep of the city.

Proposition 3. A uniquely northern housing form, based on vernacular architecture, new technology, regional climatic and topographic characteristics, and lifestyle should be developed. Apart from the vernacular architectural form of the structures developed by native peoples, there are very few examples of contemporary building forms developed specifically for the climate. Ralph Erskine and his concept of the "windscreen" building is so far the only uniquely northern design used as a principle in developing some of the new settlements in the Canadian north (Schoenauer, 1976). In less harsh northern climates such as Toronto the "international style" of corporate towers is the predominant architectural form. Glass towers in Houston are the same as those in Toronto or Ottawa. Bungalows in San Diego, California, are the same as those provided for the Inuit community in Frobisher Bay. Apartment blocks with balconies facing north, which are totally unusable for most of the year, are common sights in Ottawa, Winnipeg, and Thunder Bay!

Is it not possible to combine the historic form and experience of living in the north with new technology to create a distinct regional form responsive to the climate as well as to the style of living? An authentic regionalism should be the foremost principle in generating urban form.

In many instances our planning and zoning regulations could be blamed for the proliferation of the existing building patterns. The rigidity of existing regulations does not encourage new forms, either in individual buildings or on a neighborhood scale.

Proposition 4. Sunlight, brightness and color, and provision of private and public spaces should be considered as major design principles in the creation of urban environments. Conditions of life in northern climates are aggravated by the dearth of sunlight, isolation, lack of social interaction, deficiency in food supply and quality, and, in some cases, malnutrition.

Medical research has produced conclusive evidence of the importance of sunlight for mental and physical well-being. Sunlight counters depression, increases alertness, inhibits sleepiness, and perhaps stimulates sexual activity. The significance of those findings are far reaching (Brodey, 1985).

Obversely, there is a growing concern for the negative effects that energy-conserving measures are having on residents and workers. In the search for perfectly insulated, energy-conserving shelter, designers often forget that homes should also be delightful, bright places to live.

Office towers glamorously sheathed in tinted glass and mirrors offer interior work spaces that are in perpetual twilight.

Light deprivation experienced during the long winter months affects the majority of working adults who leave for work at dawn and return home after dark. Irritation, bad tempers, and fatigue very often result from such work patterns.

Life in the north can be very lonely as well. Distances between the few urban settlements are great. Mobility is impeded for most of the year. Isolation and loneliness appear to be major causes of alcoholism and suicide among northern populations (Siemens, 1976).

Iceland, however, can be cited as an example of a northern country in which alcoholism is not a serious problem. On the contrary, this country has a high proliferation of poets and writers who take advantage of the long northern nights to create beauty, not to destroy it.

A high level of education and a culturally cohesive society is the most probable explanation for such an exceptional lifestyle. This is reinforced by the high cost of alcoholic beverages and enforcement of prohibition measures.

Public and private spaces and places, designed to encourage social interaction and reduce the feeling of isolation and loneliness, are very important ingredients of northern communities. Climatic protection in the form of glass-enclosed streets, shopping malls, gallerias, hospitals, and nurseries are slowly becoming an integral part of the northern landscape.

Proposition 5. Urban design guidelines and principles with specific reference to winter climate and its regional characteristics are tools by which our environment can be shaped and improved. Whereas southern cities have the benefit of vegetation, color, and sunshine for extensive periods of the year, northern cities are dressed in grey, wrapped in fog, and disheveled by bitter cold winds. Expanding urbanization has created northern cities that are increasingly remote from the "natural" landscape and whose inhabitants rely increasingly on the city itself to provide relief from the dreariness of winter. The quality of life the city can offer is one of the major determinants of its prosperity—or its failure. In our battle for survival, urban design is no longer optional (Barnett, 1984).

The beauty of the built and natural environment combined with responsible civic leadership should be as important in planning our cities as land use and transportation.

METROPOLITAN TORONTO—WINTER REGION

Metropolitan Toronto encompasses 750 square kilometers of highly urbanized land along the north shore of Lake Ontario. The physical features of the region have shaped its urban growth and created the existing pattern of development. Good agricultural land situated predominantly to the west attracted the first settlers. Urban development has grown in the form of a ribbon along the shoreline because of the difficulty of bringing large volumes of water to the hinterland.

Downtown Toronto is located on a plain that slopes gently upward from Lake Ontario for a distance of approximately 5 kilometers where it reaches an old lake shoreline ridge. From this 23-meter-high ridge, the plain extends in three directions, providing the primary developable land for the metropolitan region.

Situated at 43° 40′ N parallel, its location is on par with Rome, Italy, but its climatic characteristics are similar to those of Helsinki, Finland.

The climate of southern Ontario (and the Toronto region) is modified by the Great Lakes, which raise average winter temperatures by about 3 degrees and reduce summer temperatures by about 1.5 degrees Celsius. Toronto's climate is characterized by

- moderately cold winters and moderately hot, brief summers;
- medium to heavy precipitation distributed evenly throughout the year;
- considerable sunshine, especially during summer; and
- winters with a considerable amount of snow, rain, and other precipitation as a result of cyclonic storms.

Average mean temperature in January is −4.6°C and +22.0°C in July. The coldest temperature recorded for January was −32.8°C and the highest recorded in July was +40.6°C (Environment Canada, 1982).

The mean annual snowfall of 139.2 cm is evenly distributed from December to March for approximately 41 days a year (Environment Canada, 1982). The Great Lakes influence is noticeable during the winter months, causing constant fluctuations of freezing and thawing periods leading to extremely icy conditions at various times.

Unpleasant climatic conditions are experienced when strong winds increase the windchill factor. During the winter months the prevailing wind direction is from the west; in summer the direction is reversed. Average wind speed throughout the year is 18 km/hr.

Southern Ontario receives about 2000 hours of sunshine annually. Toronto has 2045 hours. July is the sunniest month at 280 hours, and December the dullest with only 75 (Environment Canada, 1982).

Southern Ontario's outstanding agricultural productivity is to a large extent the result of its favorable climate. The Great Lakes moderating characteristics extend the growing season in autumn. The hinterland provides a firm agricultural base for this highly industrial urban area, creating a level of self-sufficiency in food supply (Brown et al., 1980).

The effect of urbanization on climate has been observed in the Toronto region from the first meteorological readings in 1840, when Toronto had 15,000 people, to the census in 1983, when the population reached 2.3 million (Brown et al., 1980). For these 143 years the amount of precipitation has increased slightly and the mean temperature in the central area has risen by about one degree.

As urban sprawl in the Toronto region reaches further into its hinterland and development replaces undeveloped lands within the central area, additional changes to the macro- and microclimate will occur. Increases in wind velocity can be expected as a result of tall office towers concentrated within very compact areas, and the "heat island" effect, fog occurrences, and high humidity levels may cause further increases in average temperatures. In the future, refreshing summer breezes from Lake Ontario that cooled the hot asphalt and announced the presence of the nearby water will be enjoyed only by sailors and those living close to the lake.

Physical and administrative tools are available already to combat some of the negative effects of climate and urban development. Table 3.1 provides some strategies aimed at improving the microclimatic characteristics through social, economic, physical, and administrative measures.

Some of these strategies have already been implemented or studied. Bus shelters and canopies at street level over subway entrances can be seen throughout most of the metropolitan area. Recreation programs during the winter season are available to seniors and school children in parks and community centers. Skating is a favorite pastime on the outdoor skating rinks at the City Hall Square and the Scarborough Civic Centre. A wind impact study for the harborfront area has produced special policies with regard to the siting of buildings, tree planting, and erection of windscreens.

But there is still much more to be done. First, the attitude toward our winter season has to be changed; once we accept the fact that we are living in a cold environment for eight months of the year, our urban form, transportation facilities, planning, and land use policies will have to reflect that reality. And then, as Walter Kehm said, "let's enjoy and make pleasurable this magic natural phenomenon called winter" (Kehm, 1985).

DEVELOPMENT STRATEGIES FOR 2001

For others outside the local planning profession it is usually difficult and confusing to differentiate between various terms, such as *Toronto, the city of Toronto, Metro, metropolitan Toronto, metropolitan region*, and so on. These terms are very often used indiscriminantly, at times referring to the political body, at other times to the geographical area.

The city of Toronto, established in 1834, is the oldest entity of the metropolitan area. The central area is a part of the city but is also considered, for planning purposes, to be the central area of metropolitan Toronto. The CBD (central business district, "downtown") is a subsection of the central area.

Metropolitan Toronto is a federation of 6 municipalities, established in 1953 for the purpose of coordinating development within its boundaries. It is also referred to as Metro or metro Toronto. The Metropolitan Toronto Planning Area coincides with the political boundaries of metropolitan Toronto, although it originally included the tier of surrounding municipalities, referred to as "the fringe," which now form part of the surrounding region. The planning area is termed a *joint planning area* because each area municipality has its own official plan and zoning bylaw (which are required to conform to the Metropolitan Official Plan).

The metropolitan region is made up of metropolitan Toronto plus the four surrounding regions: Peel, Halton, York, and Durham. Established in 1973, the growth and development of the four regions have had a strong influence on the future of the area with regard to employment distribution and housing. For that reason metropolitan Toronto policies and strategies are very closely related to the development of the surrounding regions (see Figure 3.1).

BASIC OBJECTIVES AND POLICIES OF THE PLAN

The *Official Plan for the Urban Structure* (Metropolitan Toronto Planning Area, 1983), approved in 1981, reflects present land usage,

TABLE 3.1
Strategies Aimed at Improving Microclimatic Characteristics

Toronto's Climate Characteristics	*Strategies to Counteract Negative and Maximize Positive Aspects of Winter*				
	Goals	*Social*	*Economic*	*Physical*	*Administrative*
Lengthy duration of the cold season	• Develop positive attitudes to winter • Reduce cost associated with long heating season • Reduce public expenditures on snow clearance and removal, road maintenance, and so forth	• Recreational programs • Winter festivals • Education regarding diet, health, and fitness	• Subsidies to low-income groups	• Provision of public outdoor/indoor spaces • Introduction of light and color • Provisions of recreational facilities	
Bright sunshine 2000 h/yr	• Maximize solar exposure	• Outdoor recreation programs such as cross-country skiing and skating	• Incentive programs to private sector to provide for passive solar gain and innovation	• Appropriate orientation • Passive solar design • Open space, sun pockets	• Adopt bylaws based on energy conservation criteria and siting and design for solar orientatic and solar access

Strong winds averaging 18 k/h (windchill factor)	• Reduce wind velocity	• Orientation by buildings • Building form and height • Landscaping • Walls, windgates, and protective systems	Revise bylaws and request wind impact statements
Medium to heavy precipitation: rain, snow, and freezing rain	• Ease of mobility for pedestrians and motorists • Pedestrian protection • Property protection	• Heated sidewalks • Snow removal • Nonslippery surfaces • Canopies, arcades • Heated bus shelters • Flood control	Seasonally adjusted traffic managements

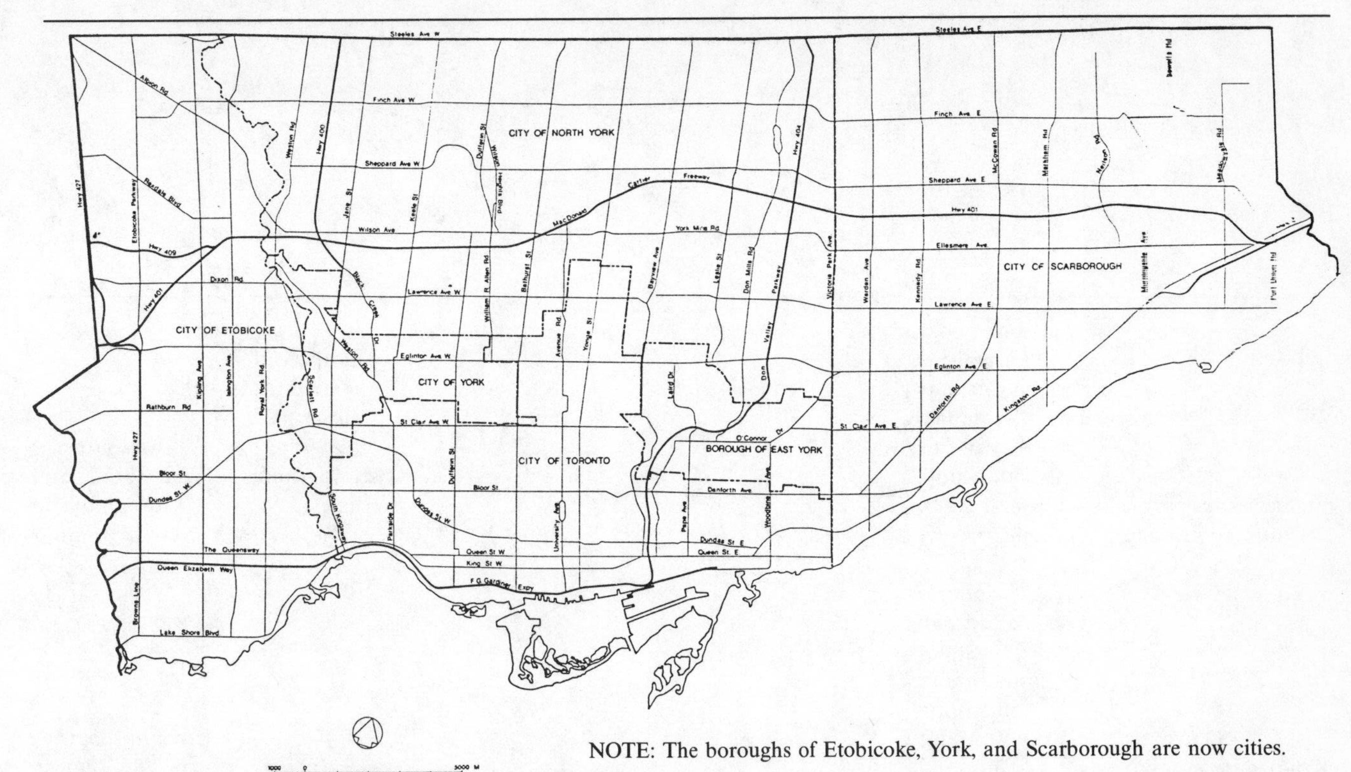

NOTE: The boroughs of Etobicoke, York, and Scarborough are now cities.

FIGURE 3.1: The Metropolitan Toronto Planning Area

economic patterns, and demographic trends. Official plan policies acknowledge the fact that the metropolitan area is almost completely developed and that emphasis has to be placed on incremental revision and modification of the urban fabric in order to utilize the existing investment in infrastructure and services.

Multicentered urban structure

The basic direction of the plan is to create a multicentered urban structure through development of six subcenters along rapid transit corridors: two within the city of Toronto and four in the suburban municipalities. The central area is to remain the preeminent business, cultural, entertainment, governmental, and management center of the metropolitan region, at the same time accommodating a strong and growing residential component.

The suburban centers act as focal points of their respective communities, and contain a wide range of "central area" type functions and employment. Their urban form is multifunctional in land use, but more pedestrian-oriented and intensive in their development relative to the surrounding areas. The six subcenters are intended to provide reasonable job alternatives to the central area for many suburban residents, primarily through the provision of office and service employment (see Figure 3.2).

The central area, as the dominant focus, remains strong and vibrant in competing for employment with the suburban centers, and in fact, it seems that actual employment figures will exceed the projections in the Metro Official Plan. Obsolete industrial land and downtown railway yards are now the prime developable areas, and in 1981 total employment in the central area was 381,770, or 69.7% of the total of the city of Toronto. Employment projections indicate that by 2001, the central area will reach a level of 453,000, or 72% of the total city level. In the office category, it is estimated that approximately 74,500 new jobs will be added between 1981 and 2001, an increase from the present 222,000 to 296,000 (Metropolitan Toronto Planning Department, 1984a).

In conjunction with strong promotion in the private sector for prestigious office space in the central area, there is also a strong public commitment for more housing units, especially in the low- and medium-priced categories. The demand is for smaller rental or ownership units conveniently related to transit and other amenities.

The Metro Official Plan has proven successful in promoting and encouraging this multicentered urban structure. The subcenters are

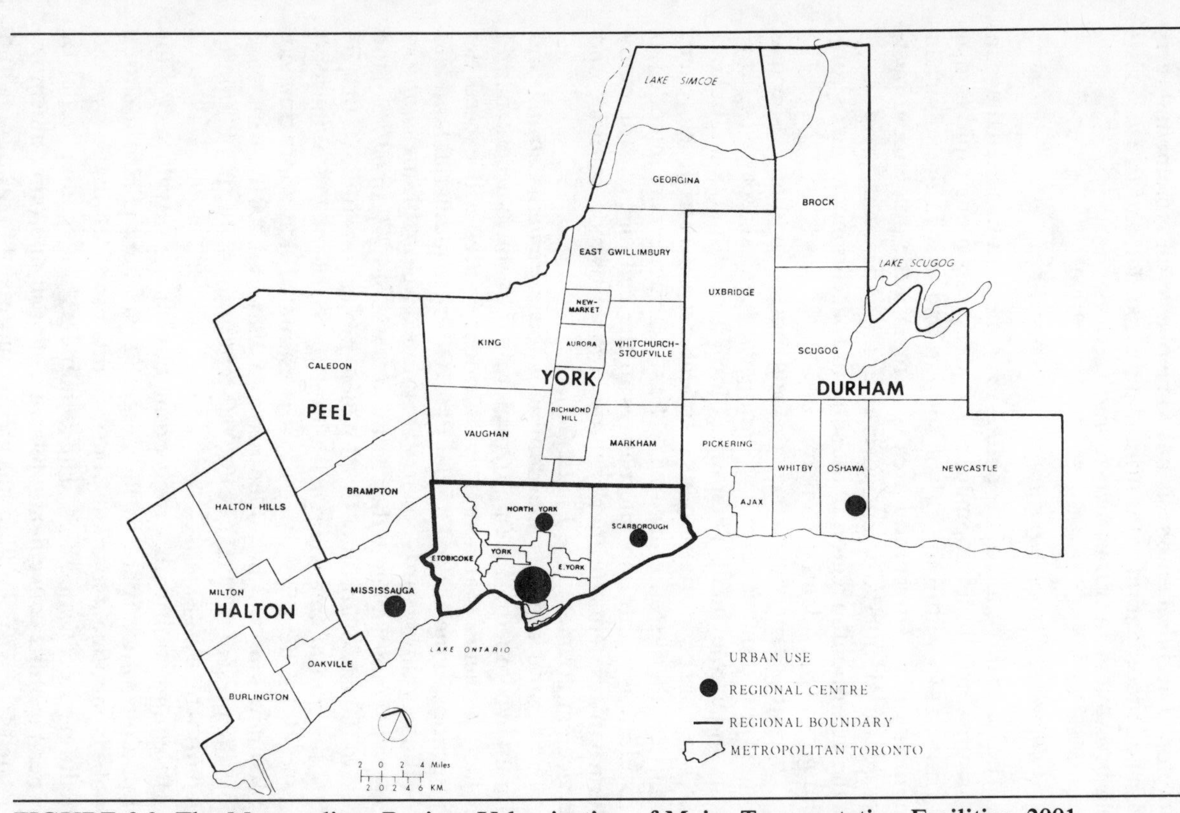

FIGURE 3.2: The Metropolitan Region: Urbanization of Major Transportation Facilities, 2001

being promoted as suburban "downtowns" where alternative lifestyles can be pursued. The major impetus for the development of these centers has been provided by various levels of government through the construction of public facilities and improved transit services. Substantial growth in office employment since 1971 has been observed.

Development of two major subcenters, the North York Civic Centre and the Scarborough Town Centre, is proceeding well—so well that it is anticipated that their projected employment targets (Metro Official Plan, Schedule 1) of 40,000 and 30,000 jobs, respectively, will have to be raised. Presently, total office space, including existing development and active applications, represents about two-thirds of the projected target for 2001. In order to compete with the central area for the same employment market these subcenters have to offer more, hence the quality of the urban environment is becoming an increasingly important tool in their marketing strategy.

Buildings of exceptional architectural design, pedestrian amenities including climatic protective devices, well-designed indoor and outdoor public spaces, and the availability of recreational and cultural facilities are becoming deciding factors for companies choosing suburban center locations.

Distribution of Employment Activities

The objectives of the Metropolitan Plan are to ensure that sufficient land will be available to accommodate the additional 240,000 new jobs expected by the year 2001. The assumption is that a majority of this growth will be in the office sector. It is expected that there will be 673,000 office jobs, or an increase of 17.2% from 485,566 in 1981. Office employment will increase its share of total employment from 39.2% in 1981 to 47.5% by 2001 (Metropolitan Toronto Planning Department, 1984a).

Out of an anticipated 188,000 new office jobs in metro Toronto by 2001, 86,000 have been allocated to the six subcenters and 70,000 to the central area.

The future distribution of new jobs will be based on concentrating the largest proportion of future employment along existing and planned transportation facilities and providing opportunities for closer home/work linkages.

Recent analysis of the employment trends in metro Toronto shows the following:

- Total employment in metro Toronto will increase from 1,240,000 in 1981 to 1,445,600 by 2001 (Metropolitan Toronto Planning Department, 1984a).
- Although much of the future population growth can be expected in the outside regions, employment opportunities are expected to grow within Metro.
- Within the metropolitan regions the manufacturing industrial sector is expected to *decrease* its share of the total employment labor force in the next 20 years, from 435,959 in 1981 to 418,300 in 2011 (Metropolitan Toronto Planning Department, 1984a).
- With a declining labor force (as the population ages) and expanding employment opportunities within metropolitan Toronto, in-commuting is projected to increase over the next 20 years (Metropolitan Toronto Planning Department, 1984b).
- Slower population growth is reflected in employment projections that are further affected by the changing structure of the populati[illegible] metropolitan Toronto, it is expected that the populations will slo[illegible]se until 2001 and then decline. This decline *within* metro Toronto will be taken up by the wider metropolitan region.

These population and employment distribution patterns will have implications for future transportation corridors and development within the central area. In view of our climate and high road construction and maintenance costs, the provision of efficient public transit should have a priority over road construction, especially if other factors such as increased densities and more compact development is to be supported (see Figure 3.3).

Housing and Residential Policies

Metropolitan responsibility is limited to the broad distribution of population and residential development and the provision of assisted housing and financial participation in municipally initiated urban renewal or improvement programs. The provision of housing is an area municipal responsibility and metro Toronto acts only in an advisory manner. In recognizing demographic trends, the changing composition of the labor force, energy conservation, and individual lifestyles the plan supports area municipal policies that would increase accommodation for seniors and a variety of low-cost rental and private housing. This increase in housing accommodation will be achieved through intensification of existing neighborhoods and job centers and through redevelopment at locations accessible by public transit.

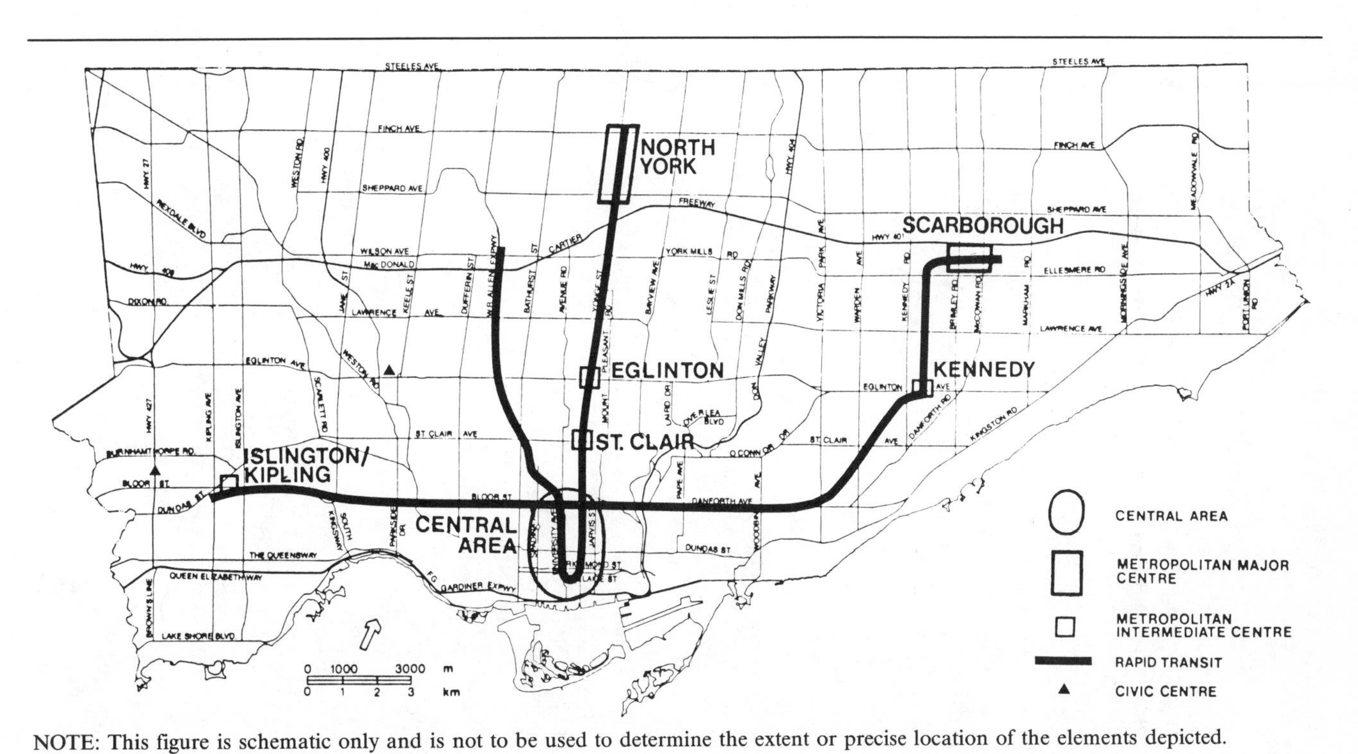

NOTE: This figure is schematic only and is not to be used to determine the extent or precise location of the elements depicted.

FIGURE 3.3: Urban Structure

Based on the birth rate, plus immigration and migration patterns from 1940 to 1960, it was anticipated that the population of metropolitan Toronto would surpass 2.8 million by 1985. The existing infrastructure built on that assumption is capable of supporting well over the existing 2.1 million. However, projections are predicting only a slight increase of 36,000 between 1981 and 2001, raising the population from 2,137,000 to only 2,170,000.

During the last three decades metropolitan Toronto has experienced the following significant shifts in its social and demographic structure:

- It is losing population to its surrounding regions. Affordable family housing and job opportunities in the regions are some of the factors causing the dispersal.
- Composition of households is changing and will continue to decline from 2.71 persons in 1981 to 2.19 persons by 2001 (Metropolitan Toronto Planning Development, 1984c).
- The population is aging and by 2001 18.7% of metro Toronto's popula tion will be over 65; an increase of 8.1% from 1981 when only 10.6% of the total population was in the 65+ age group (City of Toronto, Executive Committee, 1978).
- The rate of new household formation has been growing dramatically, especially in nonfamily households.
- There is an increased demand for the provision of modest-size ownership houses and smaller rental apartment units.
- The need for different types of community and health-care facilities will increase significantly.

The demand for family housing will remain strong and low-density residential areas will still be the preferred urban form. Because the metropolitan area is almost completely developed, new family housing will have to be accommodated in the surrounding regions. This suburban sprawl will put a great financial burden on all levels of government to provide for basic infrastructure, public transit, and social services. As public sector resources become more restricted the developer, and ultimately the buyer, will have to absorb a larger portion of the cost, forcing the price of housing to rise significantly.

Although the demand for family housing is pushing population away from metropolitan Toronto, the existing housing stock and infrastructure within metro Toronto is becoming underutilized. The loss of family households combined with fewer families with children has resulted in the closure of schools, and the reduction in household size has resulted in a decline in population and underutilization of existing houses.

SUMMARY

As previously mentioned, the Metropolitan Official Plan provides broad policy statements and directions for the growth and change of the metropolitan area for the next 20 years. In general, its *decentralization* policy expressed through the implementation of designated suburban centers, support of higher residential densities, and provision of opportunities for closer home/work relationships are in accord with many principles of development in a northern climate.

Within the context of Canadian climate and changing lifestyle there is a need for redefinition of the public indoor and outdoor space. Clearly the existing housing form and pattern of suburban residential units has to be redefined to reflect the needs of the newly emerging population in view of our climate and topography.

In harsh winter climates it is often challenging to the young and able to venture out and carry on with everyday activities, but it is nearly impossible for an elderly person to cope with the constraints of winter conditions in a remote suburban location.

During the plan review process, which is now just beginning, it is expected that concerns regarding winter livability will be incorporated within the revised document.

TORONTO'S RESPONSE TO WINTER—CLIMATE-RESPONSIVE PROJECTS

CENTRAL AREA

In order to retain the central area as a primary focus of employment, government, and entertainment the city was first to realize that the quality of life, urban form, and civic leadership were the most important factors in preserving and enhancing "livability." Creation of attractive, safe and climate-controlled pedestrian precincts coupled with the economic benefits to the city were reasons for developing the underground pedestrian system. The Eaton Centre and, soon, Harbourfront, with extensive mixed use and, residential/retail/office projects will also provide activities and spaces for all seasons.

The Underground Mall System

The largest system of its kind in the world, these 4½ kilometers of interconnected passageways and shopping malls resulted from a combination of subway construction and cooperation between the city and the private sector. What started as a modest underground linkage between four subways stops and commercial buildings has now expanded into a network, tying together 400 shops, 3000 hotel rooms, 3 million square meters of office space, and various residential projects. Between November and March you will find the streets almost empty because almost the entire work force of the downtown core is trying to eat lunch below ground! This "underground city" is constantly expanding with more and more indoor malls and benches, fountains and trees, fast food outlets, and expensive boutiques. Future megaprojects, such as the Dome Stadium, the CBC headquarters, Harbourfront, and the railway lands development will all be integrated into the system.

The Eaton Centre

Shopping malls, with their enclosed, climate-controlled pedestrian spaces and retail facilities, were the first projects responding to and providing climate protection and comfort for shoppers on a year-round basis.

The Eaton Centre is basically such a shopping center, using conventional design principles but at the same time incorporating new elements that have contributed to its enormous success. The internal shopping street was transformed into a glassed-roof galleria containing three levels of shops, restaurants, cafés, and landscaping all connected to a major subway line. In the five years since its conception, it has become a "neighborhood" street for shoppers, gossipers, pupils, peddlers, and ordinary people!

One million people pass through the Eaton Centre weekly! This project is proof of the statement made by A. Gabriel that "public spaces covered with glass can be the answer to the user's problems . . . Activity can take place there, no matter which season of the year" (Rodrigues, 1983).

During the initial stages of construction concerns were expressed by nearby shopkeepers and community groups that such a project would have a devastating effect on the Yonge Street commercial strip where the Centre is located. The "strip" at that time was already experiencing a decline. The uncertainty of retailers with regard to future development caused a rapid deterioration in the quality, type, and tenure of the shops. The street became a "sex strip" full of pornographic bookshops and massage parlors. To combat this situation policies directed toward physical improvements of the streetscape were implemented (City of Toronto, Executive Committee, 1978).

These encompassed ornamental lighting, tree planting, coordinated signage and street furniture, canopies over subway entrances, and bus and streetcar shelters. Since the adoption of these policies, joint public and private efforts have created a receptive attitude toward the improvements of the physical and economic potential of the "strip," and the Eaton Centre—previously viewed as a detriment to the street—became its biggest asset.

THE SUBURBS

Scarborough Town Centre

While the city of Toronto was developing and improving its central area, the suburbs were determined to create their own place of identity, their own downtowns.

One of the fastest growing suburban centers is the Scarborough Town Centre, serving the eastern part of Metro. The focus is its civic center, plus a major regional shopping mall, around which extensive offices are being developed. Future plans also include construction of a hotel and approximately 10,000 residential units. The development of the center is supported by the Scarborough Rapid Transit Line (RT), a modern extension of the existing subway network. A series of protected pedestrian walkways linking various buildings to the LRT station provide the same protection and comfort to the public as does the underground system in the downtown area.

Yorkdale Shopping Centre

Retrofitting of existing shopping malls to achieve more efficient use of their land area has become an increasingly attractive proposition from an economic point of view. If there is direct access to a rapid transit line, such an undertaking is even more desirable, as open parking areas can be reduced. The Yorkdale Shopping Centre, one of the largest suburban malls in Canada at the time of its construction in the early fifties, is a good example. The erection of a subway station in 1979 within walking distance of the mall made this area a very desirable location for other uses as well. In 1980 a building containing 9156 square meters of office floor space replaced part of its surface parking lot. An extensive network of climate-protected pedestrian bridges connecting all structures with the subway system is one of the project's most successful features.

Although this is a short shopping list of climate-responsive projects within metropolitan Toronto, their cumulative effect is rather significant. To quote a frequent visitor to Toronto from the sunny Mediterranean (an airline pilot who usually stays at the Downtown Sheraton Centre and extensively uses the underground system), "When they ask me how cold it is in January in Toronto, I can only tell them that it is always the same throughout the year—approximately 22°C"!

CONCLUSION

The metropolitan Toronto Area is large and diversified, reflecting the continuing evolution of its urban fabric multicultural lifestyles and complex political structure. Employment shifts, demographic trends, and lifestyles cumulatively affect its form and pattern. Metropolitan Official Plan policies were formulated with the intention of facilitating change without adversely affecting social and environmental quality.

Decentralization and the provision of effective public transit is in accord with the basic principles of building in a cold climate. Intensification of the existing residential neighborhoods will help to reduce commuting time, municipal expenses, and energy consumption, but how these policies will be interpreted at the micro level will determine the urbanity and livability of this city.

Torontonians are very proud of their Eaton Centre, quaint inner residential neighborhoods, ethnic markets and festivals, clean streets, and efficient public transit. But they are less pleased with the look of our high office towers, absence of trees, canopies, and color. They despair of the time it takes to go to work, slippery sidewalks, and bitter cold winds whipping between the concrete and glass.

So far, we Torontonians have been guilty of being followers rather than innovators, with the result that our urban environment reflects neither our climate nor our topography. Only a few projects, such as the Eaton Centre or the underground pedestrian network, are examples of what partnership between the public and private sector can achieve in accommodating our weather.

But if a more comprehensive approach toward design of our northern cities is to be pursued and more innovative projects are to be built, public attitudes have to change. We all recognize and accept the problems and aggravations of winter, but seldom are we aware of the solutions. It is much easier to go on repeating the same patterns, the same forms, using the old standards and suffering the same mistakes.

It will be up to the residents of this great metropolis to decide what kind of place they would like to live in beyond the year 2000.

Will they continue redeveloping and expanding in the same traditional way or will a very old, but "new," influence called climate be allowed to shape our environment and enhance our lifestyles?

REFERENCES

BARNETT, J. (1984) "Urban design as a survival tool." Urban Design International (Spring): 26–29.

BRODY, J. E. (1985) "Many affected by lack of sunshine." New York Times (February 15).

BROWN, D. M., G. A. McCAY, and L. J. CHAPMAN (1980) Climate of Southern Ontario. Atmospheric Environment Service.

City of Toronto, Executive Committee (1978) Yonge Street Revitalization Project, Report 32. Toronto: Author.

Environment Canada (1982) Climatic Normals 1951–80. Canadian Climate Program N. S. 1–82, Toronto.

GIBSON, B. R. (1976) "Strathcona Sound Company: lessons from preliminary planning," pp. 321–333 in N. Pressman (ed.) New Communities in Canada: Exploring Planned Environment. Ontario: University of Waterloo Press.

JACOBS, J. (1984) Cities and the Wealth of Nations. New York: Random House.

KEHM, W. (1985) "The landscape of the livable winter city," pp. 51–60 in N. Pressman (ed.) Reshaping Winter Cities: Concepts, Strategies and Trends. Ontario: University of Waterloo Press.

McLELLAN and UNDERWOOD (1983) An Introduction to Energy Conservation in Residential Development. Ottawa: CMHC.

Metropolitan Toronto Planning Department (1984a) Employment Projections. Toronto: Author.

———(1984b) 1981 Journey to Work-Travel Patterns. Toronto: Author.

———(1984c) Projections: Populations, Labor Force, Households and Housing. Toronto: Author. Municipality of Metropolitan Toronto (1986) Department of Roads and Traffic 1986 Current Estimate. Toronto: Author.

Official Plan for the Urban Structure. The Metropolitan Toronto Planning Area (1983) Toronto: Author.

RODRIGUES, G. A. (1983) A Brief Consideration of Finnish Urban Space. Department of Architecture Occasional Paper 4. Tampere University of Technology, Finland.

SCHOENAUER, N. (1976) "Fermont: a new version of the company town," pp. 316–320 in N. Pressman (ed.) New Communities in Canada: Exploring Planned Environment. Ontario: University of Waterloo Press.

SIEMENS, L. B. (1976) "Single-enterprise communities on Canada's resource frontier," pp. 227–293 in N. Pressman (ed.) New Communities in Canada: Exploring Planned Environment. Ontario: University of Waterloo Press.

4

Cities and Snow: Reflections from Finland

JORMA MÄNTY

> Helsinki is a well-ordered provincial town where it never ceases to be winter. It smells of wood-sap and oil-heating, like a village shop. Fancy restaurants put smoked reindeer tongue on the menu next to the *Tournedos Rossini* and pretend that they have come to terms with the endless lakes and forests that are buried silent and deep out there under the snow and ice. But Helsinki is just an appendix of Finland, an urban afterthought where half a million people try to forget that thousand upon thousand square miles of desolation and arctic wasteland begin only a bus-stop away [Deighton, 1976: 13].

SOME GENERAL FEATURES OF FINLAND

Finland's position "between East and West" is quite well known, but the really northern location of the country is perhaps best described by the fact that one-third of the world's population living north of the 60° latitude lives in Finland (Hustich, 1965). This third of the northern fraction of all people—4½ million—populates an area of 130,000 square miles of "arctic wasteland," indicating a rather sparse division on average, but implying that the southern parts obviously attract people and activities.

Finnish-speaking Finns are the majority, of course, but there are important minorities: Swedish-speaking Finns and Lapps with their own language.

The Finns, being of Fenno-Ugrian origin and speaking an Ural-Altaic language, have experienced Swedish and Russian influence. After having acquired independence they had to coin a cultural identity of their own, and *natural landscape*—lakes and forests, blue sky, and white snow—was taken as a self-evident symbol of this identity. However, the heritage of origins, probably in a mix with later influences, has conveyed a specific way of building up mental images, which differs remarkably from how all other Europeans build them up. Characterized in a very abbreviated way, the Finns always emphasize the *Gestalt* of the item or happening in question where as all other Europeans emphasize the movement or change of it (Strømnes, 1976). This may at least in part explain "the secret of success" of some Finnish artists and designers—Alvar Aalto being one of them. But one cannot forget the obvious effects of climate on the Finnish culture. What is characteristic of the climate in Finland is not especially the normally cold weather, but the very wide scale between warm and cold temperatures, between summer and winter, and between day and night. Top temperatures in summer are usually around +90°F and the lowest winter temperatures are around −40°F (below −50°F in January 1985). In the spring the temperatures between day and night vary from +50°F to 0°F.

SOME CHARACTERISTICS OF BUILDINGS

An obvious and self-evident fact is that to be able to live in a country like Finland one must attend to one's clothing and keep buildings and houses livable—that is, warm. All this requires energy in one form or the other. The laws of thermodynamics cannot be avoided, and thus the buildings must be insulated, not especially *against cold,* but to keep the produced heat inside. Firewood (mostly birch) was used as the main fuel in Finnish houses up to the end of fifties, but at least two general features caused a change: urbanization and the export of timber. Urbanization meant that people migrated to the south for better-paying jobs and commenced living in oil-heated blocks of flats; timber export meant rises in firewood prices to the level of oil (and later electricity) prices. This also started a nationwide change in house-heating systems. Urbanization, when seen as the extension of urban areas and as building new neighborhoods, also meant consideration of urban forms that accord with real heating infrastructures.

The modern world, having machines, electrical energy, and, especially, vehicular traffic as its characteristics, perceives ice and snow as mere nuisances. In Finland not very long ago, ice was taken from lakes, preserved in sawdust, and used for keeping food cold in summer. Snow was used as an insulating material outside cellars to prevent food from freezing. Frozen lake surfaces were used as shortcut winter roads for horse-driven sleighs as well as for automobiles. Today electricity keeps things cool in summer and warm in winter, and an enormous amount of energy is being used to keep snow and ice off the streets.

Ancient Finnish villages and the oldest towns were tightly built with wooden houses close together. In order to rationalize the parceling of village lands and to prevent devastating fires in towns, regulations were passed that from the middle of the eighteenth century onward completely restructured the Finnish rural settlement pattern (see Figure 4.1), as well as expanded town quarters by widening streets to prevent the escalation of potential fires.

One could understand very well that the original tight structures of villages and towns offered some microclimatical advantages against cold, and would do so now if the houses were made of stone or clay bricks instead of wood—rather stupid speculation, because wood was the easiest available material.

FINNISH TOWNS AND CITIES

The ancient compact—long ago extinct—villages can be said to have borne some kind of Finnish originality. But most of the towns and cities are of foreign origin, because they have been founded and largely built under Swedish or, later, under Russian rule. Only some very recently founded towns are Finnish, primarily because certain settlements, formerly classified as rural, have been converted into towns administratively. However, some new developments, or "New Towns"—Tapiola Garden City near Helsinki is an example—bear the mark of being Finnish, though with an international idea background (Rieder, 1984).

Generally, the development of any Finnish town includes extensions that largely reflect international urban structural development (see Figure 4.2).

What is interesting is that models for urban structuring and urban buildings seem to have been imported from central or rather from southern Europe without any conscious adaptation of these models to Finnish climatic factors. For example, one may wonder why certain urban elements that could have been of very good use in this cold and snowy climate—arcades and covered market passages—

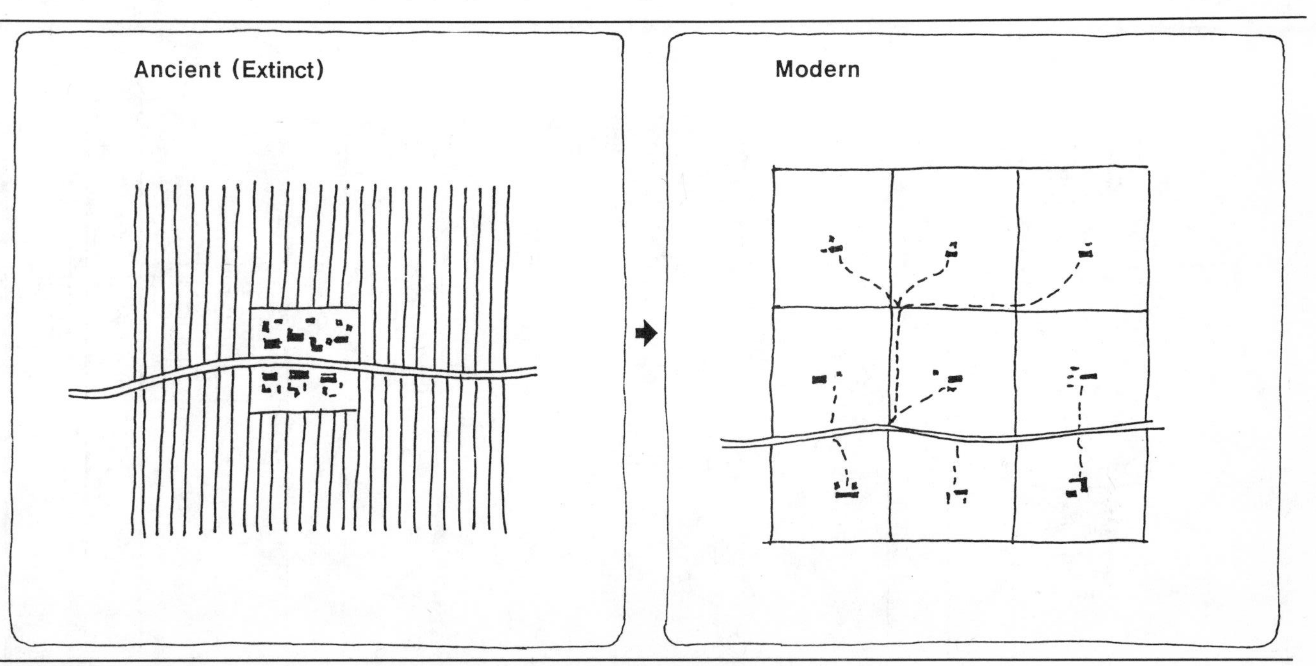

FIGURE 4.1: Schematic Presentation of Finnish Rural Settlement Pattern

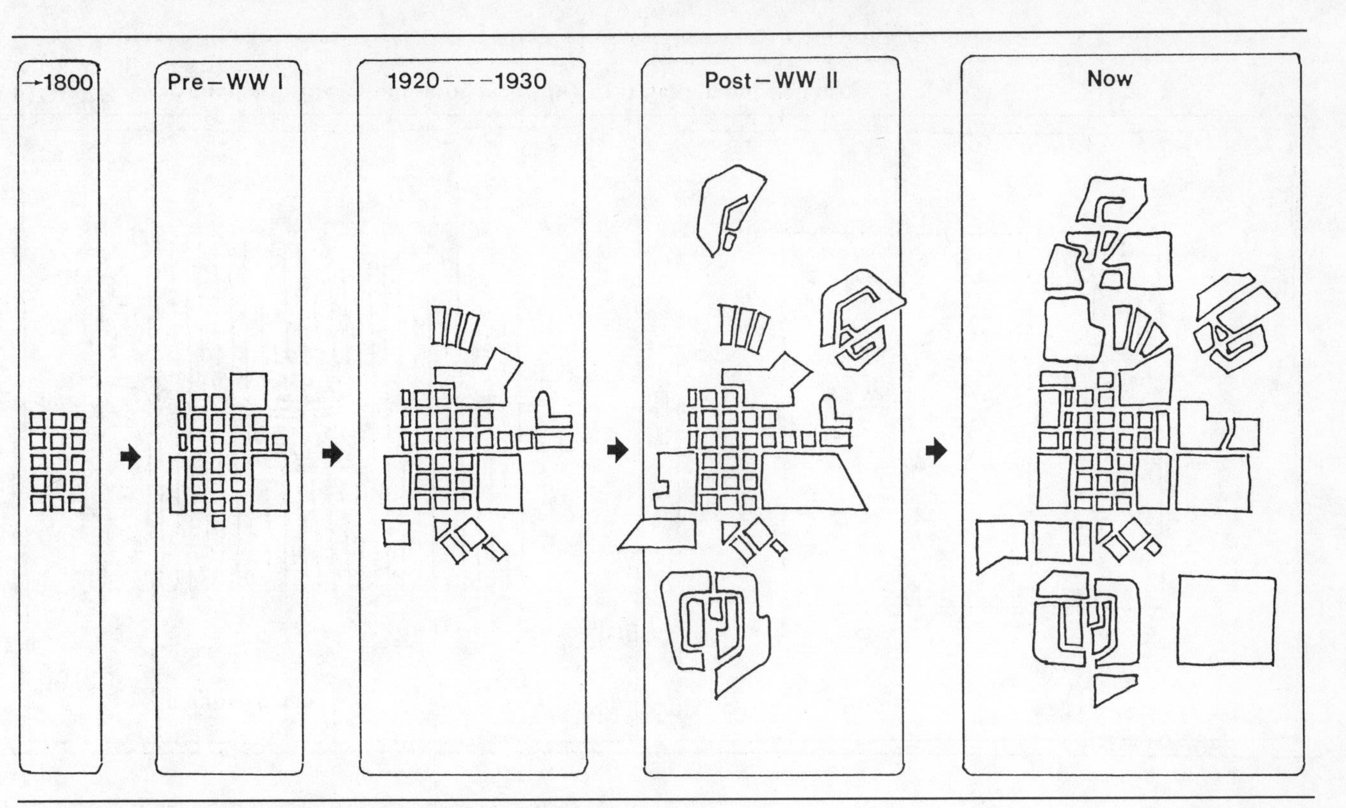

FIGURE 4.2: Typical Development of a Finnish Town

have not been adopted (not counting some really fragmentary attempts). Only very recently, planners and designers have started a discussion on the advantages of such elements.

The Garden City movement, which was known in Finland before World War II, had its first substantial effects on planning after the war, during a period of extensive building and reconstruction. More or less successful realizations of the Garden City idea took place throughout the country, the aforementioned Tapiola being internationally known. It was noticed that when buildings and streets were placed according to the topography of an area, there appeared to be no need to carry all the snow away, as has been the case in all old grid-pattern towns; ploughing the snow just off the streets and paths was enough in most cases.

The situation today is that all Finnish towns contain fragments of their stages of development, which means that the total structure of almost every town shows some incoherence. After the strong influence of functionalism (which, in fact, had its main principles written in the Finnish building law) and the Garden City idea, there seems to be at present some general indecision as to how the towns and cities should be structured. No new theory, idea, or movement with as strong an influence as functionalism seems to be circulating and at our disposal. The Garden City idea, in a somewhat molded form, seems to be still in the background of planners' thoughts.

This does not indicate that there is no search for new ideas. The energy crisis in the beginning of the seventies launched intense research activity, especially in the field of planning, in order to find solutions to save energy through some new conceptions of a rational urban structure. Very few, if any, striking results are in view. But the energy problem very clearly pointed to the need for a continuous interest in the specific climatic situation of Finland without expectations of finding ready-made solutions from countries farther south than Finland.

REFERENCES

DEIGHTON, L. (1966) Billon-Dollar Brain. London: Jonathan Cape.
HUSTICH, I. (1965) Suomi tanaan (Finland Today). Tammi, Finland.
STRØMNES, F. (1976) A New Physics of Inner Worlds. University of Tromso, ISV.
RIEDER, J. (1984) The Evolution and Development of Finnish Urbanism. Report 82, Department of Architecture, Tampere, Finland.

5

Sapporo: A Japanese Winter City

YASUO MASAI

□SAPPORO—this city in northern Japan (43°N) attracts many people. Why is that? Not many cities in northern Japan, especially Hokkaido, are very attractive to ordinary people, but there are exceptions like Sapporo. Needless to say, this large city (population 1.5 million) has many urban problems both as a great metropolis and as a winter city. The winter here (−5.5°C in January and −4.7°C in February) may not be too cold for many Europeans and Americans, descendants of ethnic groups accustomed to living in similar winter conditions for centuries and centuries. But for the Japanese it has been perceived as an extremely cold place with severe winters and with very short summers, given that the Japanese have long lived in a much warmer climate. Perception of climate can change but slowly. Only after the experience of living in such a climate for more than a century, are the Japanese—the same ethnic group living in warmer climates on Honshu, Shikoku, and Kyushu—now creating a somewhat different culture in Sapporo.

ROLE OF SAPPORO AS THE CENTER FOR DEVELOPMENT OF HOKKAIDO

Sapporo is the primary central place in Hokkaido, the second largest island of Japan. This island of 78,522 km^2 including nearby small islands, is located just to the north of Honshu. Primarily because of its cold winter it had not been settled by the Japanese until the fifteenth century. During the 250-year feudal Tokugawa period that followed, some fishermen and traders were working mostly on the coastal areas forming either permanent or temporary settlements.

The only feudal lord found on Hokkaido during the Tokugawa period was the Matsumae family, whose main castle town was located at Matsumae, a small town at the southwestern corner of the island. The Japanese were trading with the Ainu, the native people engaged in primitive fishing, hunting, and agriculture. Despite the encouragement of the Tokugawa government, development lagged so far behind that much of the land remained untouched.

After the Meiji Restoration in 1868, Japan began to develop as a modern nation in many respects. The collapse of the feudal regime necessitated people of the samurai class to find new jobs. One of the new jobs offered to them was the development of Hokkaido as a base for the militia. From 1883 to 1890 13 militia settlements called *tondenheison* were developed. Thereafter 24 tondenheison were settled, but these new ones were opened to both samurai and ordinary people. Sapporo was functioning as the government site from the very beginning, and later it rose to clear supremacy in the hierarchy of central places for Hokkaido, subjugating prosperous port towns such as Hakodate and Otaru that had always been larger in population.

Construction of the city of Sapporo was fast. The street pattern was laid out after Kyoto, an ancient Japanese city resembling Chungan from the ancient Tang dynasty of China. Its grid pattern looks quite like that of American cities—really, both follow the same principle (see Figure 5.1). The then Japanese government invited Harace Capron, the Commissioner of Agriculture of the U.S. government, as Commissioner and Advisor to the Japanese government (1871–1875). His major role was to take charge of the development of Hokkaido, particularly its agricultural, mineral, and other resources. Perhaps because of Capron, or possibly because of William S. Clark, an American from Massachusetts who taught at the Sapporo Agricultural School (1876–1877) and left among the Japanese his famous words, "Boys, be ambitious," there is a vague myth among many people that Sapporo was laid out along the line of American town planning. In fact, construction started in 1869 after the city plan by Y. Shima. Nevertheless, such a myth, true or not, can attract many—both curious intellectuals and sightseers. The relative wideness of ordinary streets in Sapporo is initially attributed to the original plan for a width of 22 m, which is significantly wider than the average Japanese traditional streets found in the "homeland." The vast expanse of wilderness, together with enlightened foresight of rosy large-scale future developments, made the construction of such wider streets

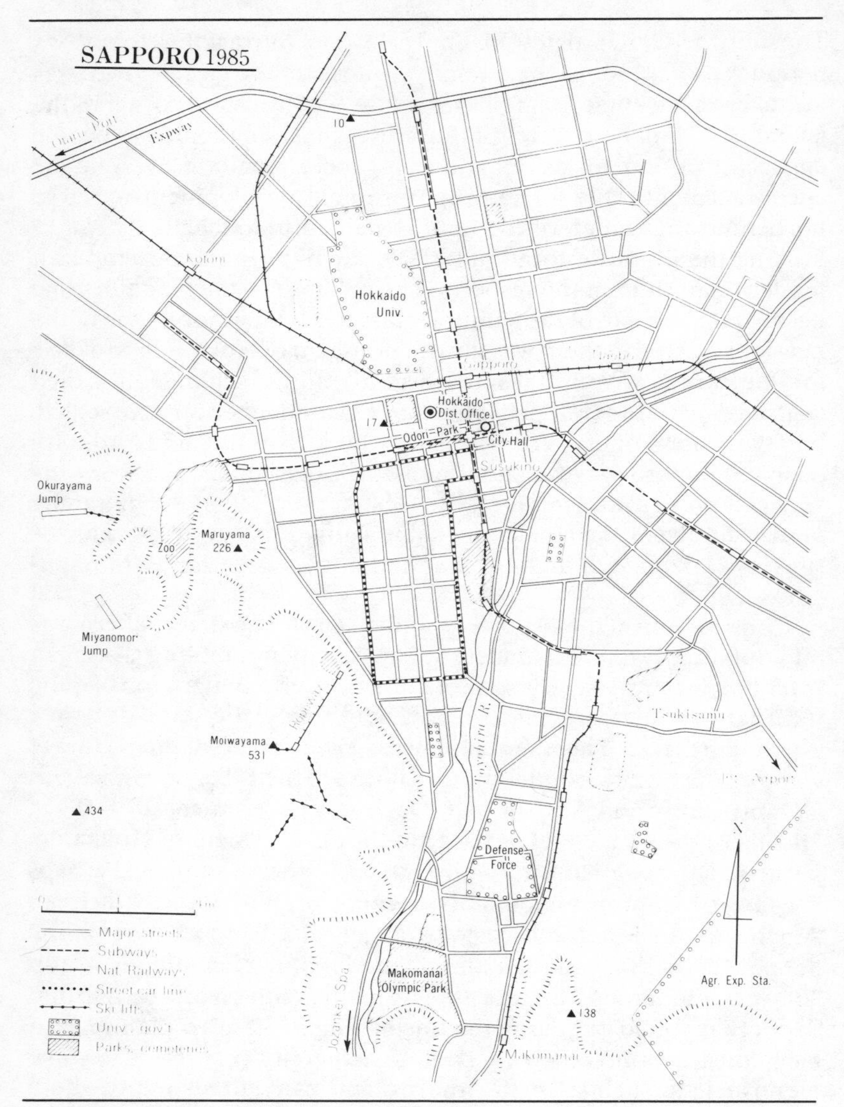

FIGURE 5.1: Map of Sapporo in 1975

possible. Some would argue that because of deep snow cover the wider streets are necessary as a logical consequence. A contrary opinion is that in the homeland, practically all the town streets are much narrower, even in the areas where snow cover is much deeper.

In any event, the spacious streets have become characteristic to Hokkaido cities and this fact is connected with the newness of Hokkaido. The Big Street or Odori (Odori Parkway), located at the heart of Sapporo, is an acme of that sort with its width of 105 m. It has culminated as a symbol of Sapporo's modernity and dignity (see Figure 5.2).

For the Japanese Hokkaido is something like the West Coast of America or Canada to North Americans. It once was a desolate pioneer fringe. It was like Siberia in that severe winter cold prevails over a vast area and it was a form of hell for exiled convicts. Hardship once was a very commonly used term for depicting this northern island. However, present-day Hokkaido is nothing like that. Although severe winter still occurs, living conditions have improved significantly. Today, many people realize that even Hokkaido, a newly developed territory, has a number of traditional or historical monuments together with easily recognizable modernity. Spaciousness once was treated as if it was something like rugged extensiveness or wilderness, but it is thought now to be grandiose or magnificent. Spaciousness is synonymous with preciousness, especially for those who live in the much congested homeland.

Sapporo was called Satporopet by the Ainu; its meaning was "dry (sat) big (poro) river (pet)." It was abbreviated by the Japanese to Sapporo, but the place name Sapporo was still quite peculiar to the Japanese for its pronunciation. It should have been written in *Katakana* (a Japanese syllabary often used for writing foreign words), but it was given Chinese characters (*kanji*) so that the people could perceive it as a Japanese place name. Most of the Ainu place names in Hokkaido were so written. It is quite certain that this endeavor, good or bad, rendered the common people's perception of Sapporo much less negative. And today, partly because of the exotic pronunciation with its double p sound, for instance, it is felt to be amusing and attractive.

PERCEPTION OF WINTER COLD AMONG THE JAPANESE

Japan stretches north to south at the middle latitudes off the gigantic land mass of Asia. The geographical condition of Japan is often misunderstood in many other countries of the world. The latitude of Tokyo is, for example, below 36° N, and it has a little snow in the

FIGURE 5.2: Odori Park in Downtown Sapporo, February 1985

winter. This may be understood by the Americans living on their East Coast, as Washington, D.C., also gets snow at the latitude of 39° N. However, for Americans living on their West Coast, Tokyo is thought to be void of snow, since even such a warm city as San Francisco is

37.5° N. The Europeans hardly realize that Tokyo gets even a little snow, since Tokyo's latitude is a little south of Algiers in North Africa. I can still remember a conversation with a Canadian women aged about 60 while eating dinner in a transcontinental train from Vancouver to the Rockies. Looking through the train window at the tall conifers along the Fraser River, she asked me, "Do you have such tall trees in Japan?' My answer was, of course, positive, in as much as Japan has similar spruces and firs, especially in the northern areas and the mountains. To my further explanation that Japan also gets much snow at the latitudes of Kansas or Virginia, she looked at me with suspicious eyes, at first not responding at all. But a few moments later, she noted, "Oh, I remember you had a Winter Olympics!" That was the Winter Olympics at Sapporo. The remaining conversation went very smoothly.

Snow has long been characteristic of Japanese culture. Many poems and works of literature have depicted snow in various forms. In many literary works, it has been treated rather "warmly." This is largely due to the fact that most of the poets or writers have lived in areas where snow does not fall much. For them, as for the average Japanese, snow had been something nostalgic, fancy, and elegant. In ancient times, the Japanese praised the beauty of snow while sitting on the floor of a house that was open to the cold wind. The traditional Japanese house had a structure designed primarily for summer heat. About 20 million Japanese live in the snow country, or Yukiguni, where snow cover is more than 50 cm and often exceeds one meter for up to six months. Even there the house structure has been quite traditional, and so their winter life has been very severe. In ancient times, the noble ladies wore a kind of kimono called *juni-hitoe*, which literally means twelve-fold kimono. In a later period, another kind of kimono primarily for men was invented, and this is called *dotera*, a very thick winter jacket made of cotton. And this winter coat, very thick indeed, was an indoor coat. (see Figure 5.3.)

House heating was minimal in the traditional Japanese house. Up until very recently the traditional house was heated only by small braziers called *hibachi* and leg-warming *kotatsu*. Room heating was negligible except for small fireplaces sunken into the middle part of the floor—*irori*. The room itself functioned partly as a chimney for the wood fires, and so the room had to be ventilated by outside cold air. Crevices and openings abounded. Frequently, sliding doors, very thin and often made of paper, were opened and the indoor temperature often was about the same as outside. Heavy clothing was then a

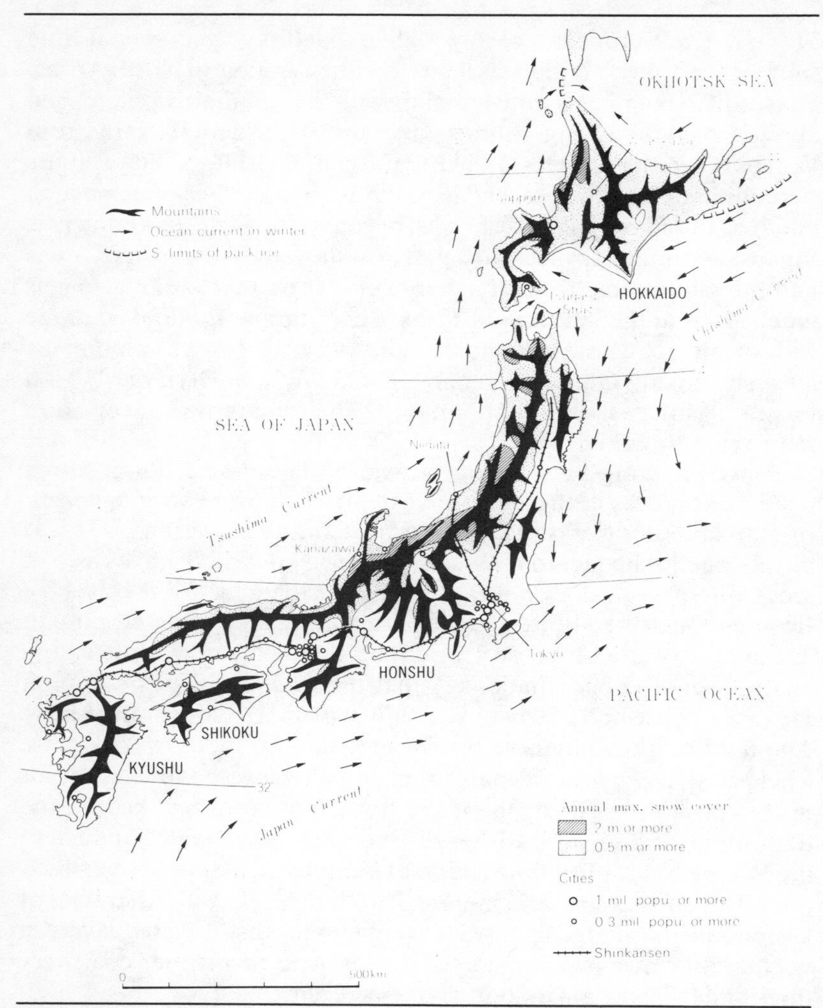

FIGURE 5.3: Winter Conditions and Cities of Japan

practical necessity. This style of life was also seen in the snow country.

The snow country was not far from most of the Japanese. As Kawabata Yasunari, a Nobel prize-winning novelist, depicted in his famous "Yukiguni (The Snow Country)," it is a beltlike region stretching all the way from western Honshu to northern Honshu facing the Sea of Japan. Hokkaido is included in the broader sense of

Yukiguni. The nearest point of Yukiguni from Tokyo is only 150 km north. In Kawabata's Yukiguni, it was several hours distant from Tokyo, as the train had to pass a very long tunnel after changing from a steam locomotive to an electric one and again to a steam engine. Geopsychologically it was different place, far away and unknown to the traveler from nearly snow-free Tokyo. Now it is only one hour by the Shinkansen train through a number of very long tunnels; there is no feeling of having traveled a great distance. This is the area, together with many other similar places facing the Sea of Japan, where the greatest efforts to overcome all different aspects of snow hazards have been made by the Japanese. Contrary to the snow conditions in Hokkaido, the snow country on Honshu is well characterized by the heavy, wet snow that falls in much warmer and more humid meteorological conditions. In other words, people in Hokkaido could not readily make use of most of the devices and life modes developed traditionally.

The geographical location of Japan's snow country, including Hokkaido, forced the people to develop ideas and media without much contact with the neighboring nations. As is well known, China exerted a profound influence upon Japanese culture, but its influence was not great as regards the measures against snow. The pivot region of ancient Chinese civilization is semiarid without much snow. By the same token, the Korean influence was not as great as one might expect. Korea as a whole is colder than Japan in winter, but its winter climate is much less humid, resulting in much less snowfall as compared with Japan's snow country, located further to the east and affected by warmer sea waters. In Korea, as in the case of northern China, snowcover rarely amounts to more than 50 cm in the densely inhabited areas. The Soviet Union's Far East is a newly developing vast area with very little population, and also with less snow cover normally, which has resulted in very little cultural contact. Even if the Soviet Union had had much more cultural contact with Japan, their experience in the frigid, continental, and subarctic conditions could not have been introduced directly to Japan, which is more marine and warmer.

As the cultural center of Hokkaido, Sapporo has experienced many ways of coping with snow and winter coldness. Know-how came first from Niigata and other places in Honshu but without much success. Other methods were introduced from America and Europe, but it took a long time for them to be adapted by ordinary people. This was largely due to the cultural and economic differences, in spite of the fact that since the Meiji Restoration the Japanese have been encouraged to assimilate many aspects of American and European culture. Therefore, of late the direct influence has come not from the neighboring countries but from far-distant places. The economic development of Japan also has made possible a sizable amount of introduction of Western cultural traits that have been designed for winter-cold climates.

One could argue that the Japanese people must hate winter cold because they are an ethnic group originating in warmer southern areas. This is not, however, such a simple question. At least three answers may be given to this question. (1) Indeed, many Japanese living in the snow country now wish to move to the Pacific coast regions where the winter climates are much drier or a little warmer. As is well known, the Pacific coast regions today are the major urbanized and industrialized areas of Japan, enabling the residents to earn higher incomes in general. Thus, it is very difficult to judge which factor, climatic or economic, is more decisive. To be sure, the people living under deep snowcover often complain about it and their intention of migrating south is often treated as if it is a matter of course. In ancient and medieval times, however, the snowy areas facing the Sea of Japan were much better developed than the warmer, less snowy Pacific side, and even today, the average living standards of the farmers there are a little better off than those on the Pacific coast. Southwestern Pacific coasts are also losing population rapidly, despite their warmer climates. In this case, migration is to the north. (2) The general trend of Japan's development has been from southwest to northeast toward Hokkaido. Although there are many exceptions, it is generally understood that western Japan is traditional and congested, eastern Japan new and rural, and northern Japan (Hokkaido) is rough and Western. Hokkaido is felt by many to be the only place in Japan where a "non-Japanese" culture can be developed. Especially among younger generations, this northern island is warmly accepted as such. And also to many Japanese who hate the terrible summer heat (one of the world's worst), Hokkaido's summer with its abundant flowers and refreshing greenery is just wonderful. (3) The modernization of Japan in the past century has directly and indirectly been related to Westernization of Japanese culture and society, resulting in aggregate in the preference of Western or Westernized life patterns found both in great metropolises and in newly developing Hokkaido. Thus, Sapporo has been regarded by many as a large city that admits both factors, and this is one of Sapporo's major attractions.

FIGURE 5.4: Yukidome or Roof Snow Stoppers in Sapporo

SOME TECHNOLOGICAL AND CULTURAL MEASURES DEALING WITH WINTER

The long and severe winter of Sapporo has not changed much as a physical or natural condition, but much has been solved by introducing new technology and ideas. The most striking postwar change in daily life perhaps has been the improvement of house structure. As mentioned above, the traditional Japanese house had a style more or less suitable for warmer, more humid climates, and such houses were built all over Hokkaido after the Meiji Restoration. Endeavors to improve house structure were repeatedly advocated by the authorities and opinion leaders, but until the postwar period, especially the time of rapid economic growth in the 1960s, public attitudes were slow to accept new trends. Now, most houses show more or less Western structural forms with smaller windows or with fewer openings to the outside, allowing the house to stay warmer. The use of more energy is now possible thanks to higher incomes, the fact of which has made life here far more comfortable.

Snow bulldozing is done extensively for most of the streets in Sapporo. Because of the relatively heavy, damp snow that often accumulates more than 50 cm a day or a night and the annual

FIGURE 5.5: Groundwater Sprinkling in the Niigata Area

maximum snow cover of more than one meter (30-year average), shoveling of snow has been and still is required of the citizens. Roof snow shoveling, which is very dangerous and expensive work, is still done by hand, except for steep-roofed houses and reinforced concrete structures. Many traditional houses are equipped with roof snow stoppers called *yukidome* (bars or hooks on a roof to stop sudden falls of accumulated damp snow, which often hurts people on the ground), but this roof style is disappearing gradually (see Figure 5.4). As far as

FIGURE 5.6: Tanukikoji Arcade Street in Downtown Sapporo

snow on the streets or roads is concerned, much is now removed by machines such as bulldozers and trucks. Snow-melting devices mostly developed in warmer snowy areas on Honshu are rarely used in Sapporo, where the temperature is too low to use them. On Honshu, snow is often melted by sprinkling with groundwater (5° to 15°C) or is dumped by hand or machines into roadside open gutters with running groundwater (*ryusetsuko*), which can melt and discharge dumped-in snow easily (see Figure 5.5). In Sapporo, however, snow removal cannot be done this way because of the lower temperature, and the trucks dump snow into nearby riverbeds. Many sidewalks in downtown Sapporo are provided with sidewalk heating systems, much to the relief of pedestrians. The systems are managed either publicly or privately.

Sapporo also has extensive underground shopping plazas and promenades. Pole Town is the largest, connecting two of the downtown subway stations, Odori and Susukino, and this is further connected with another one called Aurora Town stretching eastward from Odori Subway Station to the TV Tower at the eastern end of Odori Park. Sapporo National Railway Station is another place to have a large underground shopping plaza and walkways. A number of very extensive subterranean shopping arcades have existed in large cities with

FIGURE 5.7: Sheltered Elevated Subway Line and its Makomanai Station in Snow

the help of air conditioners, which can make them cool enough, but in Sapporo, a major reason for the installation is against winter cold. With many shops and art decorations, the underground areas serve not only railway passengers but also shoppers and tourists coming to the city center. Tanukikoji, a traditional shopping street, is a 1-km arcade, sheltering the pedestrians from snow and rain (see Figure 5.6).

The Sapporo city subway system now has two lines, the older one running north-south and the newer east-west. The trains are equipped with rubber tires. Part of the north-south line goes above ground in the form of an elevated railway. To cope with snow blizzards, the elevated section is covered by semitransparent plastic shelters for more than four kilometers connecting four stations. This sheltered elevated subway is another tourist attraction (see Figure 5.7).

Some remarks about the cultural amenities of Sapporo need to be made. The Snow Festival (*yukimatsuri*) of Sapporo held in early February, when it is coldest, now draws a great many sightseers from all over Japan and also from Southeast Asian countries such as Taiwan, Hong Kong, and Singapore. The festival's two sites, one in Odori Park in the center and the other at Makomanai to the south, boast hundreds of snow and ice sculptures and images. In 1985, for example, 143 snow and 159 ice sculptures and images, including

SOURCE: City of Sapporo.

FIGURE 5.8: 36th Snow Festival of Sapporo at Makomanai

some 40 large ones, decorated the winter city of Sapporo at the occasion of its thirty-sixth Snow Festival (see Figure 5.8).

In addition to the Snow Festival, Sapporo boasts a variety of winter sports facilities. Thanks to the nearby mountains it is quite possible for skiers to reach the mountain ski slopes within a very short time. A few of the many slopes including ski-jumps are easily visible from downtown. The Moiwayama Ski Area is, for example, situated at the western edge of the built-up area, as are the Okurayama and Miyanomori Jumps. Skiers can jump while looking down at the brightly illuminated city. Many other sports facilities include ice arenas, skating rinks, and cross-country ski routes: the latter are seldom found in warmer Honshu. Makomanai was the main site for the 1972 Winter Olympics; the area is at the southern edge of the built-up area of Sapporo, only several kilometers from the central station, and can be reached by a subway. Note that Sapporo is a metropolis with 1.5 million people. Very few metropolises of that size in the world can provide both Alpine and Nordic ski facilities within their own city limits. Not very far from the city are located many other mountain ski resorts and scenic spas, the latter of which are supported by long-standing Japanese national habit of visiting hot springs and can attract tourists all the year round.

FIGURE 5.9: Shinkansen Train at Echigo-Yuzawa in the Snow Country 150 km North of Tokyo.

Distance from the major centers of Japan to Sapporo has long been a special concern. Even as the crow flies, it is about 900 km north of Tokyo and 1100 km northeast of Osaka. In addition, it is located on an island separated from Honshu by the Tsugaru Strait. It takes four hours by ferry from Aomori, the northern rail terminal on Honshu, to Hakodate, the southern-end port of Hokkaido. Another 286 km is necessary to reach Sapporo from Hakodate by rail beyond the 739-km rail trip from Tokyo to Aomori, plus a 113-km ferry ride, altogether 1138 km. Chitose Airport is 45 to 50 minutes away by train from Sapporo Station. Passenger travel from Sapporo to Tokyo or Osaka is now mostly done by air. But there is hope of increasing the number of rail passengers. This is due to the expected extension of snow-proof Shinkansen (high-speed computer trains with a maximum operating speed of 240 km/h) from Morioka to Sapporo (see Figures 5.9 and 5.10). The greatest gap is the Tsugaru Strait, but in 1985 a 54-km tunnel between Honshu and Hokkaido began construction. Tunnels from both sides met in 1985, and completion is expected by 1987. The tunnel will run the Shinkansen and as a result running time from Tokyo to Sapporo will drop from its present 15-16 hours (Shinkansen, ordinary rail, and ferry combined) to no more than 6 hours or so. Sapporo and all other cities and towns in Hokkaido await the coming

FIGURE 5.10: Sharp Edge at the Head of Snow-Proof Shinkansen

of the Shinkansen as a symbol of the "unified Japanese Archipelago." They point to the present reality that Kyushu has long been connected by undersea tunnels and a large bridge and that even Shikoku, much smaller than Hokkaido, is now in the news with the construction now under way of three series of long bridges.

Hokkaido is a newly developed land where the immigrants were destined to create a new culture. The language spoken here is more or less the standard Japanese or the tongue of Tokyo. Especially so is Sapporo, and the citizens do not have otherwise worrisome language-related complexes. In addition, the citizens' behavior toward fashion is very much, sometimes too much, like Tokyo's, a psychological factor showing that they enjoy a metropolitan or cosmopolitan life—never again a secluded, lonesome northern frontier-town life.

REFERENCES

AONO, H. and S. BIRUKAWA [eds.] (1979) Geography of Japan, Vol. 2: Hokkaido. Tokyo: Ninumiya Shoten.

Association of Japanese Geographers (1980) Geography of Japan. Tokyo: Teikoku Shoin.

ICHIKAWA, T. (1975) A Geography of Yukiguni. Nagano: Ginga Shobo [in Japanese]. Niigata Regional Geography Study Group (1976) Snow in Niigata Prefecture. Niigata: Nojima Shuppan.

Sapporo City (1970) One Hundred Years of Sapporo. Sapporo: Author. [in Japanese]

Sapporo City Education Committee (1977) Townscapes of Sapporo. Sapporo: Hokkaido Press. [in Japanese]

Sapporo City Education Committee (1981) Historical Maps of Sapporo. Sapporo: Hokkaido Press. [in Japanese]

Sapporo Geography Circle (1980) A Town Called Sapporo at 43°N. Tokyo: Shimizu Shoin Shoten. [in Japanese]

Toyama Earth Science Association (1982) Heavy Snow (Gosetsu). Tokyo: Kokon Shoin. [in Japanese]

TREWARTHA, G.T. (1965) Japan—A Geography. Madison: University of Wisconsin Press.

6

Anchorage: Toward an Explanation of Migration to Winter Cities

KNOWLTON W. JOHNSON, C. THEODORE KOEBEL, and MICHAEL L. PRICE

□MUCH HAS AS BEEN SAID about the quest of Americans to reside in livable communities. Weinstein and Firestine (1978), for example, discussed the growing movement from metropolitan to nonmetropolitan areas. Others (e.g., Boyer and Savageau, 1981) have focused on migration patterns from the Water Belt to Parch Belt. More recently, Naisbitt (1982) contends that the large-scale migration to the South is a myth; rather the population shift is to the west and Florida. This chapter focuses on this western migration phenomenon in relation to winter cities. In particular, we are interested in population growth in Anchorage, Alaska, and if quality-of-life attributes tend to explain why people move to western Frost Belt cities.

Naisbitt (1982) in his megatrend analysis found that the fast-growing states share two similar characteristics: romantic frontier-like atmosphere and money-making opportunities. The image is almost a flashback to the days of the Old West. Americans are migrating westward toward the land of opportunity, where government regulation is frowned upon and where entrepreneurship is prized. Alaska was discussed as one of these fast-growing states.

These results provide us with a starting point for a more empirical examination of the growth patterns of Anchorage, Alaska, represent-

---

AUTHORS' NOTE: *We wish to thank Ms. Serena Partch, a former student of the lead author during his tenure at the University of Alaska, Anchorage, for her valuable assistance in collecting up-to-date information about the city of Anchorage. Special thanks also go to the mayor's office in Anchorage for its assistance on this project.*

ing one winter city in the West. Before discussing why people go to Anchorage, we will describe the city: the migration patterns and its quality of life. In sequence, the empirical analysis focuses on relationships between growth patterns and quality of life for 277 metropolitan areas reported by Boyer and Savageau (1981). Gottmann's (1983) concept of "transactional forces" is also examined as a determinant of population growth patterns. Our concluding statements center on strategies that will better position Anchorage in the future to compete for scarce resources.

ANCHORAGE AND ITS POPULATION GROWTH PATTERNS

It is true what people say—that Anchorage is only a half-hour from Alaska and that the continental United States is referred to as the "Lower 48." Another truism is that there is wilderness in the city. Moose and bears wander into residents' backyards. Salmon swim in downtown streams. But Anchorage, more than any other city in Alaska, typifies urban America. Indeed, Anchorage is urban Alaska. The municipality of Anchorage encompasses approximately 1955 square miles. The city, SMSA, and borough boundaries all coincide. In comparison, the state of Rhode Island has about 1200 square miles.

Anchorage is bordered on the west by the headwaters of Cook inlet and on the east by the Chugach Mountains. These mountains are the gateway to the Alaska interior. The downtown area is similar to other medium-sized cities with hotels, banks, bars, and other businesses. The skyline is beginning to reflect increased urbanization as buildings get taller and taller. There are also a number of large shopping centers located on the outskirts of the city. Anchorage, in the mid-1980s, is very different than the frontier town of the mid-1970s that Joe McGinniss depicted in his book, *Going to Extremes* (McGinniss, 1980).

The population growth of Anchorage has been exponential. From 1950 through 1970, Anchorage was a moderately growing community of less than 50,000 people. Its population more than tripled from 1970 to 1980, reaching 174,431 according to the 1980 census. In 1980 two of every five residents of Anchorage had lived in another state in 1975, whereas only one of five were native-born Alaskans.

Anchorage's growth has continued at a high rate since 1980. From 1980 to 1982, Anchorage grew by 29,800, or 17.1%. Net migration accounted for 21,700 of this increase, a rate of 12.4% per year, accounting for approximately 75% of the city's growth since 1980. This growth rate is one of the highest in the nation. The population for 1985 is estimated to be approaching 250,000. This is approximately 50% of the projected population for the entire state.

TOWARD AN EXPLANATION OF MIGRATION TO ANCHORAGE

Americans are, in general, nomadic. According to Lieske (1985), each year approximately 20% of all Americans change their residences, frequently to different metropolitan areas in states other than those in which they live. Why do people move? Is it for jobs and better salaries, climate and health, in search of a new start on life, a more satisfying lifestyle, or to be close to family and friends? In short, what qualities of life influence decisions to move?

QUALITY OF LIFE IN ANCHORAGE

Quality of life means different things to different people. Novak (1982) talks about the rising levels of material well-being attained in the United States of America during this century that have done much to reinforce the prevailing myth of democratic capitalism. The power of economic affluence is also said to have elevated the good life in America (Campbell, 1981). Ladd (1978) contends that quality of life is better measured by the livability of a community, the humanity of government, the purity of the environment, personal and community health, and the amenities that promote personal growth and development.

The most widely reported measurement of quality of life is that published by Rand McNally in its *Places Rated Almanac* (Boyer and Savageau: 1981, 1985). The Boyer and Savageau ratings are covered by most metropolitan dailies and the Almanac is a very popular publication throughout the United States. Other ratings are available (e.g., Liu, 1976; and Lieske, 1985); but none of these have received as much public attention as the Rand McNally ratings.

Of the SMSAs rated in 1985 by Boyer and Savageau, Anchorage, Alaska, was 242nd of 329. In the 1981 ratings Anchorage was 148th of

277 cities. Shortly following the release of the 1985 Rand McNally ratings, the *Anchorage Times* headlines read, "There are worse cities than Anchorage, 87 to be exact." The lead article on the ratings asked how Anchorage could do so poorly in the Rand McNally ratings and yet be one of eight winners—from a field of 500 contestant cities—in the All America Cities competition: an award that the city has won three times.

One reason for Anchorage's poor rating was its very existence as a "winter city." Anchorage really bombed on Rand McNally ratings for climate and terrain; only five cities in the nation were rated as having worse weather. Six criteria were used to determine mildness: (1) very hot and very cold months, (2) seasonal temperature variation, (3) heating and cooling degree days, (4) freezing days, (5) zero-degree days, and (6) 90-degree days. Unfortunately, the mildness measures did not take into consideration the fact that below-freezing temperatures in a semiarid area such as Anchorage are much more livable than these same temperatures in a humid area such as Washington, D.C. Moreover, objective measurement of climate may have little to do with residents' satisfaction with the climate (Schneider, 1975).

Anchorage also did not fare well in the ratings of housing (307th), education (300th), crime (267th), and health care and environment (235th). Housing is expensive in Anchorage, however, income levels are also high—the highest in the nation—and local and state taxes are low. Utilities are only high if the dwelling is heated by electricity. Dwellings heated by natural gas decrease the cost of residential heating by almost 50%. For example, in 1984 a townhouse with approximately 1400 square feet could be heated on an average of $22 per month. A large proportion of the housing in Anchorage has natural gas heating.

Education in Anchorage was rated average in terms of the pupil/teacher ratio; the city scored 15.55 to 1 as compared to 13.22 to 1 in Philadelphia, the city with the highest rating on education. However, with regard to local finance support and higher education amenities, it is ranked as one of the lowest in the nation. Violence is part of the Alaska's frontier heritage, and crime in Anchorage is high, but not as high as many cities of its size. The cost of health care in Anchorage is also high; hospital room rates are only second to Washington, D.C. The average daily rate of Alaskan hospital rooms is $217. Air quality is somewhat affected by carbon monoxide in the city, which is located in a bowl with water and mountains on three sides.

On a more positive side, Anchorage rated 131st of 329 SMSAs in the arts and 87th on transportation. In competition with cities like New York, being ranked in the 40th percentile in the arts is, indeed, a plus for Anchorage. The public library system and the University of Alaska campus in Anchorage provide support for the arts. Theatre, symphony orchestras, opera, and ballet are very much a part of living in Anchorage. Events are well attended with many big name performers brought in from the Lower 48 to supplement the local talent.

Despite being evaluated as having a heavy volume of traffic, transportation in Anchorage was ranked in the 26th percentile. This high ranking is primarily due to a low daily commuting time (38.7 minutes round trip) and an international airport. There are only two ways out of the city: an expressway north to Fairbanks and one to the south to the Kenia peninsula. In total, there are approximately 35 miles of expressway in the city.

Anchorage's quality of life is highlighted by its score of 33 of 329 SMSAs on recreation (10th percentile) and 14th on economics (4th percentile). The city of Anchorage is ideal for people who enjoy the outdoors. There are numerous parks and recreational areas in the city and within short driving distances. The city is proud of its bike trail system, which networks most of the city. In the winter, the trails are used for cross-country skiing, in addition to its many lighted cross-country skiing areas and two major alpine skiing resorts nearby. A new sports arena also provides residents with major league sports entertainment in hockey and basketball. One should not forget Anchorage's "Fur Rendezvous," the city's major winter festival, and the "Great Alaskan Shootout." The Fur Rondy hosts the World Championship Dog Sled Races, and the Alaskan Shootout invites major college basketball powers from the Lower 48 to participate in a preseason tournament.

The cost of living in Anchorage is high, but so is the income. An additional economic amenity is that Alaska has no income tax, and Anchorage has no sales tax. An individual can make money in Anchorage; the average household income is $51,601, the highest in the nation as compared to the national average of $33,121. Public school teachers, for example, are the highest paid in the nation: $2927 per month.

This examination of the quality of life of Anchorage, Alaska, as measured by Rand McNally shows that the city ranks low to average on some criteria and high on others. But which of the nine criteria are most important in explaining why people might migrate to Anchorage?

MIGRATION PATTERNS AND QUALITY OF LIFE

The extent to which quality of life correlates with population growth can be tested by regressing net migration on quality-of-life measures. The research reported here used 1980 to 1982 net migration (U.S. Bureau of the Census, 1984) as a percentage of the 1980 population for each metropolitan area reported in Boyer and Savageau (1981). Also from Boyer and Savageau (1981), quality-of-life measures were included as social indicators for climate and terrain, housing, health-care environment, crime, transportation, education, recreation, arts, economics, and a "cumulative score." The 1981 Rand McNally edition, which reports on data from the early and mid 1970s, was selected as the one that would have measured the quality-of-life inputs for migrants during the 1980-1982 period.

The initial analysis focuses on the correlating net migration, with the overall quality-of-life score for the 277 SMSAs under study. The cumulative score is the average ranking of each metropolis over the nine measures of quality-of-life identified above. Empirically, we found this quality of life measure completely unrelated to net migration (R = .004, not in table form). Interestingly, this is the one measure that receives the most attention in the press, but it clearly has no relationship to population growth.

Table 6.1 presents results reflecting the correlations of net migration and each of the nine quality-of-life indicators. Of the nine individual rankings, only the economics score has a strong association with migration patterns, a finding consistent with previous research (Long and Hansen, 1979; Price and Clay, 1980; Shaw, 1975). Economics (as a rank-order score) is inversely related to net migration with a bivariate correlation of −.548 and a multivariate standardized regression coefficient (BETA) of −.487. Clearly, economics is a primary correlate of migration, but it is not exclusive. Long and Hansen, (1979: 5) reported that the "most commonly expressed reasons for interstate migration are job transfer and the taking of a new job or looking for work." Employment-related reasons accounted for 59% of the interstate migration of persons. The next most important reason for migrating was "moved to be closer to relatives" (7.1%). Climate accounted for only 5% of the moves (Long and Hansen, 1979: 7).

TABLE 6.1
Quality of Life Measure and Net Migration, 1980–1982

Independent Variables	*Bivariate* R	*Multivariate* BETA	BETA²
Economics	-.548	-.487*	.237
Crime	.318	.240*	.058
Health-care environment	.113	.119**	.014
Recreation	-.074	-.113**	.013
Housing	.049	.099	.010
Arts	.050	.089	.008
Transportation	.032	.074	.005
Climate-terrain	.128	.065	.004
Education	.098	-.023	.001

* Significant at .001 level; **significant at .05 level.

Of the remaining eight rank-order measures used by Boyer and Savageau (1981), crime was the only influential correlate of migration, as reflected in the bivariate correlation and its BETA coefficient. The crime ranking is directly related to net migration, reflecting an association between higher crime rates and higher growth rates (Price and Clay, 1980). Recreation and health-care environment have statistically significant coefficients at the .05 level, but their influence is extremely weak. The direction of recreation's relationship to migration is the same as that of economics: the higher the rank (i.e., the lower the number), the higher the net migration rate. Health-care environment has the exact opposite effect. Supply of recreational facilities, particularly the private facilities that dominate the Boyer and Savageau measure (which includes restaurants, movie theaters, and bowling lanes), might be more immediately responsive to population growth than are health-care facilities.

In this analysis, the variables not associated with net migration are as important as those that are. Of particular importance is climate and terrain, which has a very weak association with net migration. Even more dramatic, the association is in the opposite direction than expected: the higher the ranking, the lower the net migration.

Overall the nine-variable equation had an R^2 of .383. A four-variable equation not reported in table form (economics, crime, health-care environment, and recreation) had nearly the identical associational power, with an R^2 of .366. This second equation also demonstrated the stability of the coefficients reported in Table 6.1.

This analysis illustrates the problem of analyzing the growth potential of the winter city, particularly Anchorage. If climate has little to do with growth, then, ceteris paribus, the future of winter cities need not be any different than the future of other cities. Anchorage is clearly a high-growth community, apparently unimpeded by its climate and abetted by its economy.

TRANSACTIONAL FORCES PRESENT IN ANCHORAGE'S ECONOMY

What is it about the economy of Anchorage that stimulates population growth? The most common response is the discovery of oil. This may be true, but there is also something else about the Anchorage economy that makes it a candidate for continued growth. That is, Anchorage is a formidable *transactional city*.

In recent years, urbanists have coined the term *transactional* to encompass the entire range of new growth activities in metropolitan areas (Corey, 1982; Gottmann, 1983). These activities are those that have emerged from the structural changes in employment occurring in advanced economies. Employment is evolving from manual labor primarily in manufacturing to work based on the use and transformation of information and other intangible products. This evolution has been described as "the post-industrial society" (Bell, 1973), "the information economy" (Abler, 1977), and more recently "the third wave" (Toffler, 1980). In all of these conceptions, the relative position of white-collar workers is becoming more dominant than that of blue-collar workers. Moreover, a new group of economic activities has emerged that collectively are identified as "quarternary" activities (Gottmann, 1961). These include the managerial and artistic functions, government, education, research, and brokerage services of all kinds.

The metropolis has become the nexus of these information-oriented functions. Although electronic communications have contributed significantly to this evolution, face-to-face interaction remains an important aspect to knowledge utilization and decision making. The city center provides the core location for these activities manifested in corporate headquarters, office complexes, and skyscrapers, supported by nearby transactional service firms that specialize in information processing and expert consultation.

Structural changes in the economy, although creating opportunity, have brought major problems and hard times to many cities, particularly in the Snow Belt, where the concentration of manufacturing employment was the highest. Plant closings and cutbacks resulted in double-digit unemployment throughout the industrial heartland. Many cities began to depopulate as displaced workers and their families relocated to other parts of the country, especially the Sun Belt (Sternlieb and Hughes, 1975).

The development of Anchorage, however, is unique in relation to that of other northern cities. Anchorage did not experience the economic and population growth based on manufacturing following World War II. Yet, as a result of its relative underdevelopment, Anchorage has not experienced the trauma associated with the more recent postindustrial changes in employment. During the 1970s and 1980s, concurrent with macro changes in the economy and while many northern cities were losing population, Anchorage has experienced tremendous population growth.

Economic opportunities are attracting migrants. Nearly half of the state's labor force reside in Anchorage. Labor force participation in Anchorage is high in relation to all metropolitan areas in the United States. For persons 16 years old and over, 76% in Anchorage were active in the labor force in 1980. The comparable rate for all metropolitan areas was 63%. Women's labor force participation is also high in the Alaskan city. This rate was 65% in Anchorage, compared to 52% for the metropolitan United States. Women constituted 45% of the city's total employment.

What kinds of opportunities are available? Anchorage is a dominantly transactional economy. Manufacturing made up only 3% of its total employment in 1980, compared to 16% for all metropolitan areas. There were more females employed in finance, insurance, and real estate (3750) than total employment in manufacturing (2450). Non-goods-producing industries employed 84% of Anchorage's labor force. Service and trade industries had comparable employment shares for Anchorage and the metropolitan United States. One of five workers in Anchorage was employed in a professional service industry. Although there is relatively low employment in manufacturing, Anchorage has relatively high employment in selected tertiary and quarternary industries, as shown in in Figure 6.1. For example, transportation, communications, and utilities employed 12% of Anchorage's work force, compared to 8% in all metropolitan areas. Likewise, public administration has a dominant 15% employment

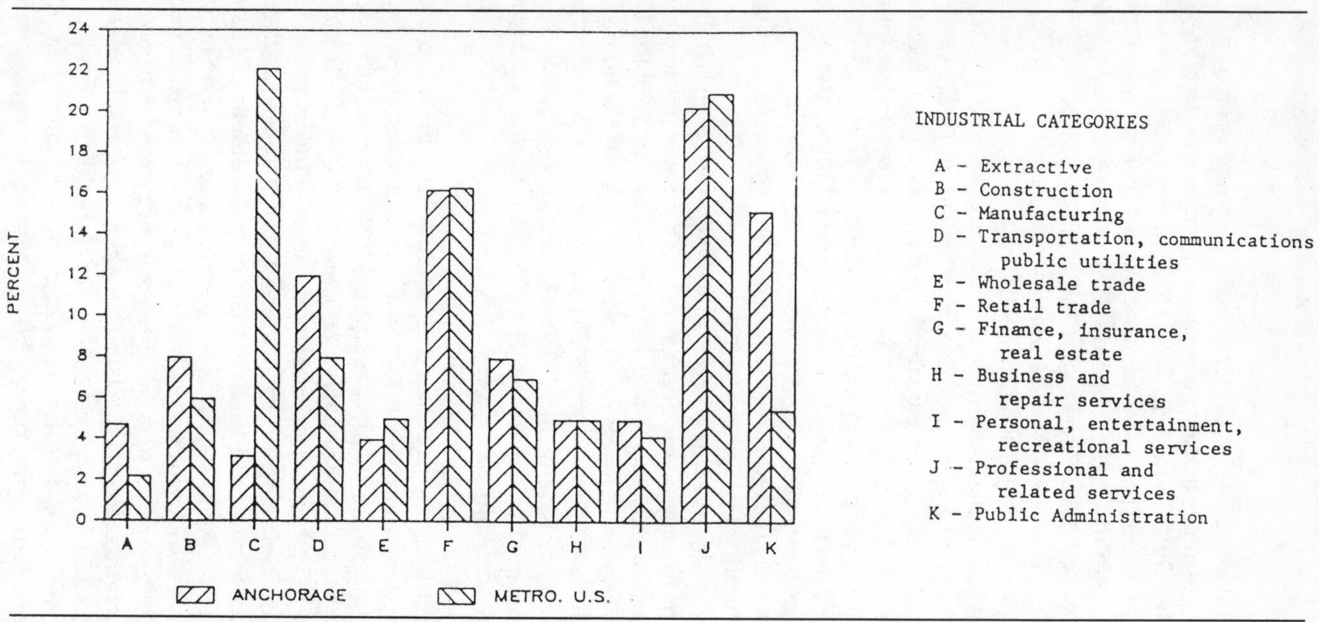

FIGURE 6.1: Employment by Industry for Anchorage, Alaska, and all U.S. Metropolitan Areas, 1980

share in Anchorage and only a 6% share in the metropolitan United States.

Anchorage's transactional economy has attracted a work force that is mobile, highly educated, and predominantly employed in professional and managerial activities. Educational attainment figures from the 1980 census show that, of the adult population in the Alaskan City, 87% had graduated from high school and 27% had attended college for at least four years. The comparable figures for the metropolitan United States were 69% and 18%, respectively. The relatively high levels of educational attainment are reflected in the type of jobs offered in the city. Of every 10 workers in Anchorage, 3 were employed in professional and managerial occupations—compared to 2 of 10 in all United States metropolitan areas. Moreover, 4 of 10 managers or professionals in the city were women.

The transactional forces in Anchorage have emerged in response to other areas of economic development, legal, and social forces. It serves as the primary brokerage center for a state vastly abundant in natural resources. Businesses and corporations involved in real estate and natural resource development, including oil exploration and extraction, have located in the city. The conservation and stewardship role of government in regard to natural resources also has created demand for many public administrators and professionals. Environmental legislation and enforcement are growing areas of concern. The rights of Native Americans and Eskimos, often involving property settlements outside Anchorage, have also contributed to the viability of the legal community in the city.

In order to meet the demand for jobs associated with these growing transactional forces, Anchorage has had to import its "human capital." Opportunities have attracted a professional, white-collar work force. However, as Anchorage's development has occurred in response to larger economic forces, the city's economic and demographic growth is vulnerable to decisions and events beyond its control.

A STRATEGY TO HARNESS ANCHORAGE'S TRANSACTIONAL FORCES

In 1984 it was announced that the oil industry in Alaska will be short lived. This reality has created a high degree of uncertainty in the state's ability to continue to offer various economic amenities that have been assumed to be part of the Alaskan heritage, such as no state taxation and bonuses for citizens. Because of the oil industry, Anchorage has acquired a far more valuable resource by becoming the information and transactional center in the state. In addition to being the primary service hub for the state, this city is also a gateway to the Orient and to Russia. It is proposed that a transactional planning strategy be developed that highlights Anchorage as a major transactional center in the global economy. Corey (1981) has called for transactional planning in Seoul, Korea, another transactional city.

Foremost, Anchorage has to support its educational system, particularly its universities. Jean Gottmann (1983: 48) contends that the university is an essential component of the modern city as a factor of fluidity in the capability of a society to adapt to the rapidity of demographic growth and technological change. This institution can stimulate information flow and transactions. There is no law school or medical school in Alaska. Also, there are limited graduate programs in Anchorage-based universities. Higher education should be instrumental in keeping youth in the state and in developing a high-caliber work force. Further, continuing education programming needs more support, as well as research services to private and public agencies. Agencies should not have to go outside of the state to obtain input for critical decision making about the future.

The state should promote tourism as a leading industry for Anchorage. Presently tourism is competing with fishing as the second largest industry following oil. The beauty of Alaska is a valuable asset that can be capitalized on in the future.

There should also be a push to move the capital from Juneau to Anchorage. In 1983 citizens of Alaska voted on whether or not to move the capital from Juneau, a small population center of approximately 12,000, to Willow, an even smaller community, where a new city could be built for an astronomical price. Centralizing government in Anchorage, Alaska's largest city, can stimulate the flow of information and transactions for the entire state and provide a more efficient location for these activities.

Finally, Anchorage should push to become a major international trade center. The city is located equidistant between many major world cities and U.S. cities on the east coast. It links the Lower 48 with the Orient and Russia.

The future of Anchorage rests in its ability to utilize the transactional forces in continuing to be the fastest-growing winter city in

America. Its primary source of power is information and those industries that produce and diffuse knowledge. The extent to which the city's leaders can simulate transactions on local, state, national, and international levels will determine the quality of life for its residents of tomorrow.

REFERENCES

ABLER, R. (1977) "The telephone and the evolution of the American metropolitan system," in I. de Sola Pool (ed.) The Social Impact of the Telephone. Cambridge, MA: MIT Press.

BELL, D. (1973) The Coming of Post-industrial Society: A Venture in Social Forecasting. New York: Basic Books.

BOYER, R. and D. SAVAGEAU (1985) Places Rated Almanac. Chicago, IL: Rand McNally.

CAMPBELL, A. (1981) The Sense of Well-Being in America. New York: McGraw-Hill.

COREY, K. E. (1981) "The transactional metropolitan paradigm: application to planning the metropolitan region of Seoul, Korea," pp. 39–49 in J. W. Frazier and B. J. Eptein (eds.) Proceedings of Applied Geography Conferences. Binghamton, NY: Department of Geography, State University of New York.

COREY, K. E. (1982) "Transactional forces and the metropolis." Ekistics 297 (November/December): 416–423

GOTTMANN, J. (1983) The Coming of the Transactional City. University of Maryland, Institute for Urban Studies, Monograph Series 2 and Occasional Papers in Geography 5, College Park, MD.

LADD, C. E. (1978) "Traditional values regnant." Public Opinion 1: 45–9.

LIESKE, J. A. (1985) The Quality of Life in U.S. Metropolitan Areas. Cleveland, OH.: College of Urban Affairs, Cleveland State University,

LIU, B. C. (1976) Quality of Life Indicators in U.S. Metropolitan Areas: A Statistical Analysis. New York: Praeger.

LONG, L. H. and K. A. HANSEN (1979) Reasons for Interstate Migration. Washington, DC: U.S. Bureau of Census, Current Population Reports, Special Studies, Series P-23; No. 81.

McGINNISS, J. (1980) Going to Extremes. New York: Knopf.

NAISBITT, J. (1982) Megatrends. New York: Warner.

NOVAK, M. (1982) "Mediating institutions: the communitarian individual in America." Public Interest 68: 3–20.

PRICE, M. L. and D. C. CLAY (1980) "Structural disturbances in rural communities: some repercussions of the migration turnaround in Michigan." Rural Sociology 45, 4: 591–607.

SCHNEIDER, M. (1978) "The quality of life in large American cities: objective and subjective social indicators." Social Indicators Research., pp. 495–509.

SHAW, R. P. (1975) Migration Theory and Fact. Regional Science Research Institute, Bibliography Series 5, Philadelphia.

STERNLIEB, G. and J.W. HUGHES (1975) Post-industrial America: Metropolitan Decline and inter-Regional Job Shifts. Rutgers University, Center for Urban Policy Research, New Brunswick, NJ.

TOFFLER, A. (1980) The Third Wave. New York: William Morrow.

WEINSTEIN, B. and R. FIRESTINE (1978) Regional Growth and Decline in the United States. New York: Praeger.

7

Minneapolis: The City That Works

DONALD M. FRASER and JANET M. HIVELY

THE M-FORM IN ACTION

In 1981, a book called "Theory Z" focused national attention on a new approach to business management adapted from the Japanese: developing the idea of cooperation among management teams. The book by William Ouchi, a management professor from UCLA, became an instant best-seller and launched a series of "new era" management books.

In 1984, Ouchi went beyond "Theory Z" to describe "The M-Form Society," in which government and business work together as a team to build a healthy community. In this latest book, Ouchi focused on the city of Minneapolis as "The M-Form in Action," where corporate and government leaders are tied closely to make up a community with a social commitment and memory.

> What is most striking about Minneapolis is not its level of corporate giving, nor its consistent ability to develop new industries, nor its ability to reorganize its industry. What is remarkable in Minneapolis is a community. It is a community of people who are connected to one another, who place peer pressure on one another, who remember for 50 or 100 years who has been helpful in the past and who has not. Like any community, it can at times be forceful, even heavy-handed in insisting on local values, but on the whole, it succeeds not by diminishing individualism, but by creating a balanced environment in which entrepreneurs. . . can build companies and create visions. Individual energies are balanced by a network of concerned peers, with an interest for the long-run health of the community. The corporate individualists, rather than being narrow-minded profit seekers at odds with their environment and adversaries of their city, become part of the community.

Ouchi was celebrating a social and institutional "memory" that supports public-private partnerships that in turn support community development.

WHY MINNEAPOLIS?

Both geography and demography have a lot to do with the creation of this city that works.

GEOGRAPHY

- The city is part of a metropolitan area that grew up isolated from other major metropolitan areas. Physical distance from major markets has fostered self-reliance, forestalled development of heavy manufacturing, and encouraged economic diversity. Physical isolation builds the sense of community.
- The city's winter climate encourages people to attend meetings rather than concentrate on more individualistic recreation. There is also nothing that can bind people in a common cause more than commiserating about below-zero temperatures or digging out after a snow storm.
- This "City of Lakes," in contrast to many other central cities around the nation, has managed to maintain a substantial upper middle-class population and strong residential tax base because of the amenity of nine lakes within its boundaries. Each of the lakes is ringed by public park land that is accessible to the full mix of city residents as well as to visitors. The physical environment is benign for development of a "livable city."

DEMOGRAPHY

- Minneapolis was settled by Yankees whose traditions included a strong work ethic and emphasis on schooling. These traditions were shared by later Scandinavian, German, and Polish immigrants. Emphasis on school and work is demonstrated by the state's first-place record for high school completion and the city's fourth highest record for labor force participation. The emphasis on education was recently demonstrated by a poll showing that the strong majority of residents would rather forego a tax cut and increase educational services. The Minneapolis public schools are organized to provide educational programs "from the cradle to the casket," and the largest student population on one American campus attends the University of Minnesota in Minneapolis/Saint Paul.

• The city's history during the 1900s has fostered participation in strong labor unions, credit unions, and housing cooperatives, and liberal politics. The former mayor of Minneapolis, Hubert H. Humphrey, pulled together the Democratic and Farmer-Labor political parties in the mid-1940s to create the present DFL (Democratic-Farmer-Labor Party), which still reflects a liberal direction. A basic trust of government supports public sector leadership.

• The egalitarian spirit dominating Midwest politics has fostered a system of decision making that is open to anyone interested in participating—whether it's in the biennial precinct caucuses that elect delegates to political conventions or meetings in the city's 84 neighborhoods. Newcomers are welcome as long as they are willing to show up at meetings.

• Pluralism is the key descriptor for Minneapolis. It has neither a "weak" nor "strong" mayor nor City Council. The system is collegial, operated through committees bridging multiple boards of elected officials. Nothing gets done without a consensus; in order to resolve conflict there must be active and continuous communication and compromise in Minneapolis.

OVERLAPPING RELATIONSHIPS

Overall, the Minneapolis community exists in a web of overlapping relationships that brings together in close contact the business, civic, political, and other sectors. These contacts take place within dozens of committees, councils, and other private groups organized by concerned citizens.

In his article, "Community Conflict," the sociologist James Coleman describes the importance of overlapping membership in civic organizations to resolution of community conflicts. In Minneapolis many people are active members in several kinds of organizations simultaneously, siding with one interest group in one setting and against them in another, bridging the gaps between groups of people advocating specific interests.

Ouchi describes several of the unique groups that entice participation by business managers. Among them are the following:

- The *Downtown Council*—responsible for reversing the trend toward inner city decay in the 1960s and creating, along with local government, the Nicollet Mall. Along the spine of the mall lie shops, offices, and new downtown housing. Connected to the spine is a network of skywalks—

enclosed pedestrian bridges—linking 32 city blocks and fostering a secondary retail system that protects pedestrians and shoppers from inclement weather.

- The *Minnesota Business Partnership*—a group of senior managers who participate in task forces studying public matters "affecting not only their own business interests but also the interests of Minnesota and its citizens."
- The *Minneapolis Project on Corporate Responsibility*—joined by five dozen corporations and devoted to better integrating corporate and civic objectives.
- The *Citizens League*—a collection of 3000 individual and 600 corporate members who participate in studies and follow-up implementation of recommendations regarding major public issues as widely disparate as vocational education, taxes, neighborhood organization, and cable communications.
- The *Urban Coalition*—created by business leaders responding in 1967 to minority outcries for inclusion in the social and economic mainstream of city life.

In these civic organizations as well as in the more traditional Chamber of Commerce or religious and social organizations, business leaders participate in a network in which peer pressures tend to transform narrow self-interest into a broader and more enlightened dedication to community interest. The emphasis on broad participation in a rational and accessible decision-making process is not limited to business people, however. Many thousands of the city's 365,000 residents are similarly active in neighborhood and school and park or other organizations attacking commonly perceived problems. Few public decisions are made without a lot of discussion by community organizations.

This participatory ethic was reflected by the recent comments of a native son. During his campaign for president, Walter Mondale kept interrupting his speeches to ask the question, "Don't you agree?" This was taken by some as a sign of weakness. In Minneapolis, it's an expected part of the process.

THE "PASSING OF REMOTENESS"

Harlan Cleveland, director of the Humphrey Institute of Public Affairs, talks about "the passing of remoteness." The economic revolution created by development of computers and telecommunications has led to the demise of isolation. At one time, geographic location was critical. Economic future depended on location—on a river, on a coast, at the intersection of trade routes. Now, information is more critical than location. Networks of interest are more important than geographical communities. Proximity is less dominant as a factor whereas new communities are formed based on interests, skills, and knowledge.

The question is how the sense of community in Minneapolis, couched in regional isolation, will be affected as we progress further into the Information Age.

On the one hand, Minneapolis should be in a good position to take economic advantage of this change. Its highly skilled, well educated, work-oriented resident population should provide a solid base for development of technology-based industries. A new partnership involving the city, university, and business community has initiated the "Minnesota Technology Corridor" between the University of Minnesota and downtown. Both a concept and a specific location, the Technology Corridor will foster growth of information-based industries.

Other groups sponsored by local government are making plans to cultivate the media industry and to assure an infrastructure that will assure the city's leadership in the telecommunications industry.

The question still remains, however, about how the "social memory" in Minneapolis will expand or change with the demise of its distinct remoteness.

THE AGING CENTRAL CITY

Minneapolis is not immune to the cycle of aging Northern cities:

- Although the outmigration of middle-class families to the suburbs common in the early seventies has been reduced in the eighties, it is still occurring.
- Although the city's healthy economy and conservative budget process still support a triple-A credit rating, local resources cannot keep up with the need for repair of aging housing stock and deteriorating infrastructure.
- Competition from the lower-taxed suburbs, rural Minnesota, adjacent states, and the Sun Belt presses hard on the ability to maintain a healthy economy in this aging Northern city.
- Immigration of dislocated workers from depressed agricultural areas combined with an influx of poor people, including large

numbers of refugees, is continuously contributing to an increasingly cosmopolitan complexion of residents.

- The old guard of home-grown CEOs is changing and the social memory is subject to attrition as professional executives primarily interested in bottom-line profits come to Minneapolis as one step on the corporate career ladder.
- Most important, the growth in jobs is not aiding an increasing population of very low-income single parents to break out of poverty and become economically self-sufficient.

In Minneapolis just as in other central cities, the rich are getting richer and the poor are getting poorer. The gap between the rich and poor is growing, and the problems associated with poverty—child abuse and neglect, anxiety and mental illness, and chemical dependency, petty theft, and violent crimes, as well as apathy—are growing.

As cleavages between social classes grow, as the society becomes two tiered, the shared sense of community is diminished—and with it what makes Minneapolis special.

CLOSING THE GAP

In spite of the impact of the communications revolution, local self-reliance must still be the watchword for progress in Minneapolis. Adequate aid to accommodate the municipal overburden is not forthcoming from either the federal or state government. Minneapolis must take advantage of its strongest asset—its sense of community responsibility—to assure the possibility for participation by all of its residents in the "good life."

Minneapolis Mayor Donald M. Fraser has initiated a variety of public-private partnerships to achieve this goal. Examples follow:

- The *Minneapolis Community Business Employment Alliance*, called "McBEA," brings business, labor, and community representatives together to work out strategies to aid the hard to employ. The group is currently focusing on community-based business enterprises, in which hard-to-employ workers with entry-level skills receive close supervision while performing relatively simple functions such as book binding and scavenging recyclable materials.
- A *Neighborhood Employment Network* matches a lead business with a social service agency in each inner-city neighborhood to work with hard-to-employ residents on job counseling, training, and placement. City employment and training programs are run through this decentralized administration.

• A *First Source* program requires all businesses receiving public assistance through tax-exempt financing, land cost write-downs, and so on to consider hard-to-employ Minneapolis residents first for all entry-level jobs.

• A *Youth Work Internship* program guarantees private sector summer jobs to low-income potential dropouts between their junior and senior years, and full-time jobs to these participants following high school graduation.

• A new *Youth Coordinating Board* made up of elected officials from the city council, park board, library board, board of education, and county and state legislature has been formed to monitor youth development in general. Implementation of a strategy to address problems of adolescent pregnancy and parenting will be high on the board's agenda. Another strong interest is in assuring appropriate early childhood development for low-income, educationally disadvantaged children. A task force identified a strong relationship between early childhood development and later employability.

• A *Project Self-Sufficiency Program* is coordinating services such as housing assistance, child care, health care, emergency loans, and legal aid to support very-low-income single parents to pursue individualized plans to achieve economic self-sufficiency.

• A *Transitional Work Internship Program* has been developed to place youth who have graduated from high school but are still unemployed nine months later into one-year internships in city departments.

• A *Greater Minneapolis Labor-Management Council* with equal participation from business and labor is working to mitigate the harmful effects of business-labor conflicts as well as to develop a cooperative stance in planning for the economic future.

Mayor Fraser recognizes that the problem is not a lack of available jobs—the official unemployment rate in Minneapolis is under 5%—but rather assuring placement of hard-to-employ Minneapolis residents in these jobs, as well as assuring the support services needed by single parents seeking economic self-sufficiency.

The basic objective is to make sure that everyone in Minneapolis faces the problems of an aging city experiencing the divisive economic pressures placed on all central cities in America and shares in the social and economic benefits of this community.

Perhaps the most difficult task is to convince business leaders that cultivating the economic self-sufficiency of hard-to-employ residents is not simply a charitable act but essential to long-term survival for the Minneapolis community.

Minneapolis has been known as the city that works because of its cohesive sense of community. How that sense of community survives the "passing of remoteness" and divisive class differences spawned by the information-based service economy is worth watching.

Part II

The Transactional City: Issues and Cases

□Almost every mature city in the United States was developed in the nineteenth century during the Industrial Revolution and with the opening of the Western frontier. In most cases the history of each of these cities is based upon its transportation and locational advantages or upon its marriage of immigrant labor to new manufacturing establishments. Much of urban history in North America has been uniquely related to the development of manufacturing employment. In urban America the billowing smokestack was a sign if progress and prosperity for almost a century.

In his clever book, *The Nine Nations of North America*, Garreau (1981) characterizes in particular the cities around the Great Lakes (including the Canadian side) and the middle Atlantic states as "the foundry" region. As he writes:

> Cities are the Foundry's dominant physical characteristic. There are lots of them.

They're not terribly far apart, by the standards of most of the continent, and they are crowded places [Garreau, 1981: 66].

But Garreau also writes:

> The problem with the Foundry is that it is failing. Its cities are old and creaking, as is much of its industry. It is still struggling with its historic role as the integrator of wildly different personalities and culture and ethnic groups, and there is no assurance that the sociological battles that it has been assigned will end in victory [Garreau, 1981: 65].

His characterization extends to Canada as well—especially Southern Ontario—beginning with Windsor, which is actually south of Detroit, and reaching up to Ottawa and the mill towns along the St. Lawrence river.

In this region the urban history is a history of North American manufacturing and the image of that history is strongly oriented to the invention of machinery. Therefore, in facing the problems of industrial decline and the identification of future prospects, most of the solutions proposed are concerned primarily with the restoration of manufacturing employment. But now, because of deindustrialization, as the business cycle ticks back up in the late 1980s the buffalo (e.g., well-paid blue-collar jobs) are not returning to the industrial valleys of the middle of North America. Instead, in cities such as Pittsburgh and Baltimore, a transactional economy has evolved that is achieving new gains in the emerging service or information economy. And as was indicated in the previous section Anchorage — the fastest growing city in the United States—has only 3% of its total employment in manufacturing. Urban growth is no longer driven by manufacturing.

THE INDUSTRIAL NATURE OF THE SERVICE ECONOMY

Recently, the executive director of the Coalition of Service Industries, Inc. (Sims, 1985) has outlined a particular interpretation of the so-called service economy:

> Since the 1940s, the United States has had more of its citizens employed in the business of providing services than in the production of goods. That makes us a service economy.
>
> But the two go hand in hand. It is helpful to think of it in terms of automobile production. Once a car rolls off the assembly line, services make it valuable—transporting it to the dealer, advertising it for sale, financing the loan, insuring it against damage or theft, maintaining it in working order.

Sims continues:

> These services may be more or less expensive and expedient than they were 20 years ago. But the point is not that there are more service workers, so services should be cheaper and faster. The point is that

there are more consumers, our transactions are far more sophisticated, and our equipment continues to be perfected and technologically advanced.

So America's service economy is humming along. But we need new trade negotiations to reduce barriers to trade in services abroad; we would like tax reform to give favorable treatment to service firms, and we would like our Government to collect better data on the service sector.

In conclusion Sims indicates the following:

Seventy-four percent of the U.S. work force is employed in the service sector. It generates 69 percent of our gross national product and keeps up a surplus in the services balance of trade.

The partial point in her analysis is that much of the service sector does not provide personal services that contribute to the productivity of the manufacturing sector. Industrial productivity is a direct function of the quality of services provided to industrial establishments.

Another insightful interpretation about the changing nature of manufacturing was provided recently in *Forbes* magazine (Cook, 1982). Cook relates the "downsizing" of U.S. energy consumption to the "downsizing" of the American automobile and to the "downsizing" of other industrial products that supply the automobile industry. He reports that General Motors is already halfway through a program to cut vehicle weight in half by 1995. That program will sharply reduce the amount of iron and steel, labor, energy, rubber, and glass required by each automobile.

Cook projects a similar downsizing in other American industries, guided by the new capacities of the microprocessor. The microchip is to industry what the hybird seed was to agriculture. It permitted an expolsion of production of agricultural products simultaneously with a dramatic decline in agricultural employment.

As Cook writes:

Looking at it in this way, the very advance of electronics, of electronic data processing, dooms the older, labor-and-capital-intensive heavy industries. They are still needed but are no longer center stage. Hence the analog with the old agriculture holds true: The knowledge industries are replacing the blue-collar industries in the advanced societies as the service and blue-collar industries once supplanted agriculture as the principal activity of mankind [Cook, 1982: 165].

The same article quotes an industrial leader, Reuben Mettler (TRW, Cleveland):

> The exciting thing thats going on is the use of the new technology in the older and more mature industries. What you can do for a combine, a tractor, a truck or a car with a little electronics is pretty dramatic. It enhances quality, it enhances productivity, it reduces costs and it permits more sophisticated designs. [Cook, 1982: 165].

But these advances in industrial productivity have coincided with the expansion of the service and information economies. From 1970 to the early 1980s total employment in the United States rose by 25 million new jobs, from 75 million in 1970 to over 100 million by 1983, but only 2.3 million were in manufacturing. Virtually all the new jobs have come in the service industries: utilities and transportation, wholesale and retail trade, producer services, and government and personal services. And the expansion of this service sector has been most pronounced in those cities that have ave, one way or another, been able to make the transition to transactional centers.

Actual manufacturing has been transferred to the so-called "greenfield factories" in Tennessee, Iowa, and almost everywhere else where there is cheap land and cheap labor. A sign of the times was an announcement in July 1985 by General Motors that their new Saturn plant was going to be located in bucolic Spring Hill, Tennessee, once known as the "male capital of the world." This latest locational decision perhaps symbolizes the reality that new manufacturing employment is not going to be necessarily urban based, in either the Sun Belt or the Snow Belt. But the knowledge and technical services that design the factories of the future are likely still to be located in the centers of industrial know-how—Detroit, Cleveland, Akron, and so forth, especially those with linkage to the expanding global economy.

In the last decade, the global economy (in terms of measured GNP) has grown by a factor of three, but world trade has increased by a factor of seven. Here again the expansion, involving multinational and international complexities, has been facilitated by electronic technologies and organizational relationships that are inherent in the concept of the transactional economy.

Although the argument can be made that the healthy expansion of the global economy requires transactional cities, it is also apparent that not every winter city can successfully aspire to be an effective transactional center. But the concept must be introduced and elaborated into the vocabulary of development of cities in northern latitudes. This section represents a modest step in that direction.

SECTION OUTLINE

In the first chapter in this section Corey reports and reviews the status of the transactional city paradigm. He interprets and extends the work of Jean Gottmann and illustrates how he applied the concept of transactional urbanization to the preparation of a master plan for the future of Seoul, an Asian winter city.

Another of Gottmann's notable concepts was that of the megalopolis. Mayer reviews that idea and reports on the status of the Great Lakes megalopolis that extends up the St. Lawrence river to Montreal and Quebec. Because of the manufacturing and industrial capabilities associated with this megalopolis, it differs in many respects from the megalopolis of the Boston to Washington corridor with its preeminent role in communications, culture, and politics.

Berry and his colleagues elaborate the new status of Pittsburgh, which in three decades has transformed itself from a manufacturing center to an unusual transactional city. Its recent ranking in *Places Rated* as the number-one city in terms of livability has attracted new attention to this city and its unique characteristics.

In his chapter Knight presents a view of industrial cities that are learning to appreciate the role of knowledge as a factor of production. In his view the industrial know-how of manufacturing cities remains even after the manufacturing employment has left. In his framework the information or service economy is driven by the knowledge that workers and cities must create the conditions and environments that are going to retain or attract such talent. In the unrelenting competition for talent, cities must create a cosmopolitan environment of a world-class character.

Hendon and Shaw report on the role of the arts in urban development. In their view the arts are primarily winter activities that have a significant function in the redevelopment of older industrial cities through the provision of an amenity infrastructure.

Van Til and Gabel discuss the emergence of food policies for cities in northern latitudes. The Cornucopia Project of Rodale Press has pointed out that U.S. food supply lines are overextended.They advocate the development of more regional markets by winter cities, with subsequent gains in food freshness, nutritional value, taste, economic security, and reduction of transportation costs. In one sense their effort reminds us to retain an appreciation of the agricultural hinterland in the development of the concept of transactional cities as a framework for assessing the future of winter cities.

For some there might be an apparent paradox in the fact that this section opens with a concern for the nature of transactional cities in a global economy and ends with an admonition to develop a concern for local food markets. But the paradox simply recalls the slogan of the 1980 World Future Society conference in Toronto, which was "Through the '80s: Thinking Globally, Acting Locally."

REFERENCES

COOK, J. (1982) "The molting of America." Forbes (November 22): 161–170.

GARREAU, J. (1982) The Nine Nations of North America. Boston: Houghton-Mifflin.

SIMS, M. (1985) "The service you get in a service economy." New York Times (July 2) (See also Coalition of Service Industries, Inc., Washington, DC).

8

The Status of the Transactional Metropolitan Paradigm

KENNETH E. COREY

□THE ROLE OF this chapter is to provide a review of the evolution of the concept of the transactional metropolis. The approach used in this essay is to (1) state the Transactional Metropolitan Paradigm; (2) introduce the significant transactional writings of Jean Gottmann; (3) list a sampling of research and trends since the initial statement of the paradigm; and (4) end with a call for the development of a paradigm of transactional metropolitan planning.

THE TRANSACTIONAL METROPOLITAN PARADIGM

During the spring of 1980 I first stated the Transactional Metropolitan Paradigm (Corey, 1980). With minor modification since originally defined, the most recent statement of the Transactional Metropolitan Paradigm is that

> the transactional metropolis is driven by the structural evolution in *employment* from manual labor to work on tangible products where the relative position of white-collar workers increasingly is more dominant than that of blue-collar workers. The *central district* of the metropolis is dominated by special abstract, information-oriented functions that operate in *offices* and in *skyscrapers* that form *skylines*. The activities of the *city center* include the concentration of the highest-

AUTHOR'S NOTE: *I am grateful to Professor Jean Gottmann for his comments and suggestions on a draft of this chapter. His continued support and contribution to the advancement of our knowledge about transactional cities is appreciated.*

> level forms of customized decision making and knowledge utilization, as in the management of multi-national service forms that are specialized in the processing of information and the provision of expert consultation. Both face-to-face contact, and electronic communications are central to the effective operation of all these transactional establishments in the metropolitan core. Other elements of transactional centrality include visitors and *transients* who come to the central district primarily to transact business, and also to take advantage of the amenities and dynamic activities of the metropolitan "crossroad." The constant presence of these transactional transients, in turn, shapes and confers vitality to the region, especially to the metropolitan center. Universities, cultural institutions and recreational opportunities are particularly reinforcing of transactional behavior.
>
> The *outlying areas* of the larger modern metropolis increasingly receive employment in manufacturing production, wholesaling, branch offices, and retailing that services nearby residential parts of the suburbs. These suburban firms cluster in *subcenters* on major links in the metropolitan *transportation network* that provide accessibility both to surrounding low-density residential areas, and to the center of the metropolis.
>
> The transactional metropolis is connected to, and interwoven with other transactional centers, forming *metropolitan networks* within national territories and across international boundaries. The principal transactional metropolitan *concepts* developed by Jean Gottmann include: *terms of employment, hosting environment, interweaving of quaternary activities, evolution of urban centrality, the Alexandrine Model*, and *Megalopolis* [Corey, 1984b: 77].

This statement is a synthesis of findings and interpretations principally from a selection of Jean Gottmann's publications to 1980 (Corey, 1980). The would-be paradigm statement was formulated to aid in the planning of a metropolitan development strategy for Seoul, South Korea, in the year 2000. Before describing that and related efforts, it is necessary to elaborate on the work of Professor Gottmann that formed the basis of the Transactional Metropolitan Paradigm.

JEAN GOTTMANN AND THE TRANSACTIONAL CITY

The "father" of the transactional city concept is Jean Gottmann. For thirty years or so Professor Gottmann has been researching and writing about the modernization of cities and city regions. His coverage of the topic and the significance of his conclusions have been rich and insightful. From his Virginia research in the 1950s (Gottmann, 1955) to his monumental 1961 book, *Megalopolis: The Urbanized Northeastern Seaboard of the United States*, to his numerous 1960s and 1970s articles on quaternary economic activities and with his book, *The Coming of the Transactional City*, in 1983, few other scholars have made such a valuable contribution to our understanding of the recent forces of urbanization occurring in the advanced economies of Western Europe, North America, and Japan. We are fortunate that he has compiled a comprehensive bibliography of his more than 250 publications (Gottmann, 1983b). With this listing one can review his important observations and conclusions about the evolution of the city into the transactional metropolis.

Gottmann first introduced the transactional concept in print in 1960 in *Megalopolis*.

> The growing mass of goods and of data to be taken into account in the operation of business aimed at markets of Megalopolitan size requires much space and more and more personnel. Servicing modern production and consumption requires handling the goods on the one hand and the transactions on the other. Managing the transactions is no longer a simple matter of counting and contracting, or arithmetic and legal forms; today it involves information and research on the technology of products or of management and public relations, as well as on the events in the markets, local or distant information that cannot be obtained and used efficiently without education, competence, and special skills. Indeed, one wonders whether a new distinction should not be introduced in all the mass of nonproduction employment: a differentiation between *tertiary* services—transportation, trade in the simpler sense of direct sales, maintenance, and personal services—and a new and distinct *quaternary* family of economic activities—services that involve transactions, analysis, research, or decision-making, and also education and government. Such quaternary types require more intellectual training and responsibility. The numerical increase in the tertiary and white-collar jobs appears to be related to an accompanying rise in the number of the professions and specializations classified under these older labels [Gottmann, 1961; 576–577].

His first paper on the transactional concept was published in 1971 in French (Gottmann, 1971). Gottmann reflected recently that "between 1961 and 1971 the idea was gradually elaborated in several of my articles and many lectures" (Personal communication, May 10, 1985).

Some of the main elements of Jean Gottmann's concept of transactional urban dynamics have included the following:

(1) Terms of employment. These are the conditions that set "the time spent at work, during the day, the week, the year, the place of work, the nature of the services, the duration of the arrangement, the remuneration and benefits received by the employee" (Gottmann, 1978: 393). Changes in the terms of employment permit more time and income for commuting, multiple places of residence, entertainment and continuing education, and so on; thereby contributing to the areal spread, and the lowering of the density gradient of the transactional metropolis.

(2) Hosting environment. This is the complex of amenities and the quality of life factors that are attractive to transactional actors (Gottmann, 1979: 5). These factors enhance the centrality of the center city and operate to service the transactional transient (e.g., conference goers or business persons from out of town) and metropolitan resident alike.

(3) Interweaving of quaternary activities. These are white-collar occupations and functions characterized by accessibility, information flows, transactional performance, quality of the labor market, amenities and entertainment, expert consultation, money and credit markets, specialized shopping, and educational facilities (Gottmann, 1970: 329–330). Resulting transactional webs operate within and between metropolitan areas, both nationally and internationally.

(4) Evolution of urban centrality. Urban centrality "consists of a multiplicity of central functions gathered in one urban place; and it rests on one or several networks of means of access converging on that place" (Gottmann,1975: 220). In preindustrial periods the central functions consisted of *administration* in the form of the *castle*, *commerce* in the form of the *market*, and collective religious *rituals* in the form of the *temple*. Later, urban centralities followed cycles of production, first being concentrated in cities (i.e., fourteenth to seventeenth centuries); second, being scattered (i.e., late seventeenth to nineteenth centuries); third, concentrated again in the early twentieth century; and since the 1920s production again is dispersing while transactions are dominating center-city employment (Gottmann, 1975: 224).

Gottmann's other major notions, "The Alexandrine Model," "Megalopolis," "hinge," and "crossroads" also have contributed to the transactional paradigm. They are described in other publications (Gottmann, 1961, 1972).

EVOLUTION OF THE TRANSACTIONAL METROPOLITAN PARADIGM

The evolution of the Transactional Metropolitan Paradigm is reviewed in three sets of developments: my work on Seoul, South Korea; the related transactional activities at the University of Maryland; and transactional contributions most recently from Jean Gottmann and others.

SEOUL AND THE TRANSACTIONAL METROPOLITAN PARADIGM

After reviewing much of Gottmann's publications on transactional cities (Corey, 1980), I found it surprising and inexplicable that so few urban scholars and urban planners seemed to have known about or utilized the transactional city concept in their work. Despite this omission, I found the concept to be informative and useful in proposing a future development strategy for the metropolitan region of Seoul (Corey, 1981). The transactional metropolitan planning strategy called for a series of actions designed to reduce Seoul's contemporary problems and to capitalize on Seoul's transactional potential. Empirical data on Seoul and the Program Planning Model (PPM) were used as the principal method to plan for Seoul's future development policy (Van de Ven and Koenig, 1976). The seven-element stages of PPM were used to develop a transactional urban policy for Seoul as follows:

PPM Stage 1: the initial mandate. With the clients and sponsors being the metropolitan government of Seoul and the Korea Planners Association, a project was initiated to develop and present an overall strategy that could be used in planning for Seoul in the year 2000 (Corey, 1982). I was the initial planner in this context. The PPM framework was used to make the planning task manageable. PPM was applied both to research the issues and its approach was recommended to the clients in Korea to be used in the actual implementation of the strategy for Seoul.

PPM Stage 2: problem exploration. The clients took the lead in specifying the problems and in providing the initial data and information on Seoul's needs. I used independent documentary sources to corroborate and complement the provided material; this resulted in a comprehensive problem exploration. Empirically, the problems of Seoul included large and rapid population growth, water and air pollution, the need for land use and transportation patterns that minimize energy consumption, and the ability to finance Seoul's new programs that are to be planned to address these problems. The needs assessment also revealed a number of transactional opportunities that required attention and planning. For example, the potential inherent in Korea's recent economic transformation in manufacturing industries would seem to offer future new options in transactional activities to the policy planners of Seoul.

PPM Stage 3: knowledge exploration. This PPM stage was executed by conducting an extensive literature review and content analysis of more than 25 years of published research by Jean Gottmann (Gottmann, 1983b). This approach was taken because it was hypothesized that Seoul's future might benefit from a long-term strategy, one component of which would be based on developing the regional economy's service sector and services-employment potential. Special attention was paid to information-age, post-industrial, and transactional employment generation.

PPM Stage 4: proposal development. The product of this PPM stage was a strategic-level proposal for Seoul's regional policy planners to consider. It was sketched out at a level of overview that was intended to provide food for thought, but was also general enough to encourage the Seoul policy planners later to elaborate and "Koreanize" the strategy into operational detail.

PPM Stage 5: program design. In order to provide the clients with examples of some of the specificity required for the implementation of a transactional planning strategy, several alternative programs were introduced (Corey, 1981). These programs included the following: tactics to avoid creation of new manufacturing jobs in Seoul, initiatives to develop clusters of transactional employment and office centers, the integration of Seoul's new and developing subway system into this strategy, the creative use of high-rise buildings, emphasis on compact spatial patterns for such development, complementary residential-area locations, capitalizing on Seoul's primacy and not dismantling its transactional potential, minimizing private automobile ownership, promotion of amenities and services for transactional transients, and so on. These were proposed as illustrative actions for ultimate tactical programming by the Seoul-region planning staff.

PPM Stage 6: program implementation and program transfer. The basic notion proposed here was to execute the strategy's programs from the principle that "risks are greatest by starting too big, not too small" (Delbecq, 1978). thus, implementation included piloting, testing, and experimenting with Seoul's transactional metropolitan strategy initially in the central business district and at selected subway crossroads in the suburbs. With the ultimate establishment of a demonstrated transactional employment base, it was suggested that lower level information-oriented services be diffused systematically to Seoul's periphery and down Korea's urban hierarchy to medium and smaller cities throughout the country. This, and the PPM substages of pilot to demonstration to full implementation experimental sequencings, followed from the propositions of PPM (Delbecq, 1978).

Program evaluation: spans PPM stages 4 through 6. The major emphasis in this PPM element was to stress the identification and use of empirical measurement criteria upon which monitoring and assessment of the planned programs of the transactional metropolitan strategy could occur and be used to effect midcourse corrections.

Numerous, simple nominal-scaled transactional indicators were proposed to measure progress toward the planning targets that were to be set for Seoul and its region. Since suggesting these early and relatively primitive empirical measures for transactional urbanization (Corey, 1980: 59–60), Nagashima has demonstrated the use of analogous, but to my view, more effective, empirical criteria in measuring Japan's Tokaido Megalopolis (Nagashima, 1981).

The evaluation program also included the recommendation to reform the Korea census as it applies to the capital region of Seoul (see Gottmann, 1983a), so as to have the census categories and variables compatible with a regular survey of these new transactional and recent information-age economic dynamics and demographics. Areal units, location of census respondents, and the periodicity of transactional behavior were made explicit and included in the recommendation for this evaluation component (Corey, 1985).

UNIVERSITY OF MARYLAND, COLLEGE PARK

Faculty research. Based on a convergence of the research of individual faculty and a recent attempt to focus research attention on the forces of transactional urbanization, the Department of Geography and the Institute for Urban Studies at the University of Maryland at College Park have been hosting a number of transactional scholarly activities. Robert A. Harper, a long-time associate of Jean Gottmann (Gottmann and Harper, 1967), recently wrote a chapter on metropolitan transactions and transactional flows, including their measurement, and people flows, paper transactions, money transactions, government transactions, and abstract transactions (Harper, 1982a). In addition, he has called for an applied transactional geography (Harper, 1982b). The book, *Modern Metropolitan Systems*, edited by Charles M. Christian and Robert A. Harper, includes other chapters by Maryland faculty that contribute to our evolving understanding of the contemporary metropolis (see chapters by Thompson, Brodsky, and Christian). Institute for Urban Studies faculty member Marie Howland has been using Dun and Bradstreet data to compare plant closings between the Sun Belt and the Frost Belt (Howland, 1984). She has concluded that there is a higher rate of plant closings in the Sun Belt. It was found that branch plants provide the least stable employment, and single plant firms provide the most stable employment. She has concluded that "economic development officials should adjust their development programs to favor local entrepreneurship" (Howland, 1984: 18).

Student research. Several recent University of Maryland graduate students also have been exploring selected dynamics of transactional urbanization. Geographer Christopher Fuchs has identified the driving forces behind private sector office-space developers in the Washington, D.C., metropolitan area as they made their development plans and development-location decisions (Fuchs, 1983). Urban studies student Martin Anderson is completing an experimental research project designed to test the Transactional Metropolitan Paradigm in the San Francisco Bay region, including the Silicon Valley area (Anderson, 1985).

Visiting faculty. Visiting geography faculty to the University of Maryland, College Park, also have contributed importantly to our thinking about the transactional metropolis. Australian-based senior lecturer John V. Langdale has investigated information-age technologies at the metropolitan scale (Langdale, 1982) and at international scales (Langdale, 1984a, 1984b, 1985). Aharon Kellerman, a senior lecturer from Israel, has published a number of transactionally relevant pieces, including a paper on telecommunications and their impact on metropolitan areas (Kellerman, 1984), and a comparative study on the transformation into service economies of the United States, Canada, Great Britain, France, West Germany, Italy, and Japan (Kellerman, 1985). He also has in preparation a paper on the unique structure of the Israeli service economy and the roles of the capital and largest cities in the evolution of Israel's services (Kellerman, 1986). Michael Smout, a geography professor from South Africa, recently examined "the way in which urban form and function may evolve in response to societal and technological change" in the postindustrial city (Smout, 1983). In this paper, he also specifies empirical issues that should be of concern to students of the postindustrial city.

Lefrak lectures. In order to complement some of the internal University of Maryland faculty interests in transactional urbanization, the recently initiated Samuel J. Lefrak Lecture Series in Urban Studies has been steered toward transactional-city themes. Jean Gottmann was the first such lecturer in 1982. He rewrote his Lefrak presentations and they were published as *The Coming of the Transactional City* (Gottmann, 1983b). The 1984 Lefrak Lecturer was Everett Rogers. His lectures addressed the Silicon Valley phenomenon; they were published as *The High Technology of Silicon Valley* (Rogers, 1985). Future Lefrak Lecturers also will be selected for their contributions toward understanding and planning the transactional metropolis and economy.

A recent direct spin-off from the Lefrak Lectures by Professor Gottmann was a proposal by John H. McKoy, former director of the District of Columbia's Office of Planning, to use the transactional city concept as a framework to recommend public intervention strategies for Washington, D.C. He saw the need for a "research project to explore the extent of 'transactional' activities in Washington, D.C., and to access their impact on the future economy of the City" (McKoy, 1984: 1). When conducted within the context of the city's policies and objectives for the land-use element and the economy element of the recently developed "Comprehensive Plan for the National Capital," such transactional-city research was seen as being useful in developing urban planning programs. McKoy observed that the District of Columbia's recent employment pattern has seen strong growth in the private sector, with federal government job production declining. District of Columbia employment has declined in construction, wholesale and retail trade, and manufacturing industries, whereas service-sector jobs have increased by 52,000 between 1970 and 1982. He noted employment gains in health services (without a significant increase in the number of hospital beds); computer-related occupations; and "transactional" activities, such as legal services, finance, headquarters' functions, and communications-related occupations. Toward the end of developing a transactional strategy for the District of Columbia, McKoy called for the following issues (among others) to be addressed:

- Evaluate the District of Columbia's *competitive employment* position relative to its suburbs and the Baltimore area in attracting and retaining jobs, especially in those categories just listed.
- Identify the current *land availability* and land use constraints on industrial growth.
- What can the District of Columbia do to develop *jobs and revenue for the city*? Can "hosting environment" jobs in hotels, restaurants, entertainment, quality retailing, and transportation services provide opportunities for less skilled workers?
- Identify the extent to which *structural employment shifts* toward a "highly technical, mechanized and transactional economy may cause structural job dislocations for significant segments of the District's and the region's labor force and public policies and programs to be followed . . . to deal with them" (McKoy, 1984: 3).

These are the kinds of transactional metropolitan research and policy-planning needs that demand attention, not only for Washington, D.C., but for cities of all sizes and locations.

OTHER RECENT CONTRIBUTIONS

When attempting to formulate the Transactional Metropolitan Paradigm in 1980, I felt that there was relatively little support available in the development-planning and geographical literature—with the exceptions of the rich writings of Jean Gottmann, and a few others such as Ronald Abler, Robert Cohen, Peter Hall, Richard Meier, Allan Pred, Jon Van Til, Melvin Webber, and some of the office-space researchers (e.g., Gerald Manners). There were available also the more generic scholars and journalists, such as Daniel Bell and Eli Ginzberg, and Alvin Toffler, respectively (Corey, 1980).

Now, five years later, I am pleased to observe that here has been an explosion of research and popular writing on postindustrial and information-age forces and transactional metropolitan patterns, planning, and management.

Jean Gottmann. Of course, since early 1980, Gottmann has continued to contribute to the development of the transactional city concept. His 1983 book, *The Coming of the Transactional City*, is one of the most comprehensive of his recent transactional publications (Gottmann, 1983b). In addition, it includes the complete bibliographic citations of his works up to 1983. He retired from the University of Oxford in the fall of 1983. In his honor, John Patten edited *The Expanding City: Essays in Honour of Professor H. Jean Gottmann* (1983). Gottmann continues to be quite active through writing and further publication, personal correspondence, and by receiving professional visitors to his home in Oxford. Recently, he reminded me of several research issues that seem important to the advancement of our collective knowledge about the transactional metropolis (personal communication, June 26, 1985): (1) Transactional forces seem to be impacting *cities of all sizes*, and not just the largest metropolises. He noted that large and small cities alike seek to attract *convention centers and related facilities* (Guenther, 1985; Koenig, 1984). For example, in Akron old grain silos have been converted into a modern hotel thereby enhancing the city's amenities and increasing its conference capability. Some medium and small cities have been studied from a transactional perspective (e.g., Edinburgh, Scotland; Oxford, England; and Nantes, France; see Gottmann, 1979), but much more transactional urban analysis is needed among the range of city sizes. (2) We need to study universities and their special role in the transactional metropolitan economy and the development of the transactional metropolis (see Gottmann, 1983b; 47–57). (3) Despite all of the advances in telecommunications, *personal contacts* remain important in transactional behavior. This relationship needs to be monitored and reassessed continually as we seek to understand the locational and spatial-organizational impacts of such technology on urban employment and urban structure. (4) Finally, Gottmann stressed the importance of research on the *empirical reality* of the transactional metropolis. Our stock of substantive knowledge needs to be deeper and more widespread so as to facilitate the integration and comparison of transactional urban findings.

Other urbanists. Just after the 1980 presentation of the Transactional Metropolitan Paradigm, Harvey Perloff's book, *Planning the Post-Industrial City*, was published. The book represents a transitional statement that seeks to link the old physical master-planning approach with newer approaches that require more attention to short-term policies planning and nonphysical concerns. Perloff's approach incorporated empirical case material from six U.S. cities, and it sought to take into account the transformations associated with the increasing employment shifts into the service sector. Michael Teitz has published an outstanding review of the Perloff book; the review is useful for our purposes in that it specifies the required attributes of a new paradigm (Teitz, 1981).

Also in 1980, geographers Stanley Brunn and James Wheeler edited a volume that examines the new dynamics of growth in the American metropolitan system. It offers a national-level perspective

(Brunn and Wheeler, 1980). Catherine Nagashima has presented suggestions for measuring empirical urbanization trends associated with the information society. Using such differentiations as daytime population patterns and nighttime population patterns, she demonstrates the use of transactional measures in Japan's Tokaido Megalopolis (Nagashima, 1981).

Columbia University's Conservation of Human Resources Project, directed by Eli Ginzberg, has produced some important research of use to the student and planner of the transactional metropolis. Two noteworthy examples include *Services: The New Economy* (Stanback et. al., 1981) and *The Economic Transformation of American Cities* (Noyelle and Stanback, 1984).

In 1982, Gary Gappert and Richard Knight contributed to our understanding of future urban change by compiling and editing *Cities in the 21st Century* (Gappert and Knight, 1982). Also in 1982, British geographer Peter Daniels published the book *Service Industries: Growth and Location* (Daniels, 1982). It introduces university students to the locational attributes of service industries in the contemporary economy of urban areas and regions. Another recent text is David Clark's *Post-Industrial America* (Clark, 1985). It includes a chapter on the post-industrial city.

A special symposium issue of the *Journal of the American Planning Association* was published in 1984 on high technology and economic development planning. It includes four articles on this relationship. The symposium editors observed that their effort

> represents the start of a discussion we think will continue for many years. One component of an economic development strategy could be technology-based. But the precise roles of different levels of technology must be defined and evaluated, not just left to develop haphazardly. This call for more research is no academic subterfuge but an expression of frustration over the limited knowledge on which we base our planning [de Bettencourt et al., 1984: 261].

High-technology development also was addressed in a recent report by the United States Congress. It includes an appendix entitled "High-Technology Location and Regional Development: The Theoretical Base" and was written by geographers Howard Stafford and John Rees (*Technology, Innovation, and Regional Economic Development*, 1984).

Dennis Gale has made an important contribution with *Neighborhood Revitalization and the Postindustrial City: A Multinational*

Perspective. It is one of the relatively few works that focuses on the transactional metropolis at the subcity scale. Neighborhood revitalization is revised in Canada, Austrailia, England, and Western Europe (Gale, 1984).

A 1983 book by British policy analysts Ken Young and Liz Mills addresses the process of urban public policy in the transactional city. Using cases from British cities, the nature of intervention and the management of post-industrial change are developed (Young and Mills, 1983). 1983 also saw the publication of an overview article by Michael Conzen that describes the major agents of change within contemporary U.S. metropolitan areas. It is entitled, "American Cities in Profound Transition: The New City Geography of the 1980s" (Conzen, 1983).

Peter Hall and his associates Amy Glasmeier and Ann Markusen have contributed to the definition of metropolitan "high technology industries" (Glasmeier et al., 1984). Hall also has written a number of recent papers that address the roots and the future of cities (Hall 1981, 1984, 1985). These pieces are useful for offering context and forecasts for the transactional metropolis. In 1985 Hall and Markusen brought together some of their own work and that of others in the edited book *Silicon Landscapes*. It treats high-technology job creation in the United States and Britain (Hall and Markusen, 1985). Castells, Hall, and others have written essays on technological change and its role in the evolution of spatial forms (Castells, 1985).

Recently, British geographer D. J. Dwyer discussed the possible application of the transactional city concept to selected cities in newly industrializing countries. He sees Hong Kong as having such potential (Dwyer, 1985). Based upon the previously mentioned planning experiment in Seoul (Corey, 1981), Dwyer's proposition might well be considered by urban researchers and urban planners for utilization and experimentation in the transactional sectors of urban areas in other developing countries (Corey, 1985).

Recent generic literature. Since 1980, a great deal of general social and economic literature has been published that is useful in providing context for the student of the transactional city. Examples include *Megatrends* (Naisbitt, 1982) and *Cities and the Wealth of Nations: Principles of Economic Life* (Jacobs, 1984).

It should be noted that transactional forces are not restricted to North America and Western Europe. The economies of Japan, South Korea, Taiwan, Hong Kong, and Singapore also should be monitored and researched (Burks, 1984; Hofheinz and Calder, 1982; Masuda, 1980).

TOWARD A PARADIGM OF TRANSACTIONAL METROPOLITAN PLANNING

This essay concludes with the observation that a theory of transactional metropolitan *planning* is needed. Such a theory would inform planning practice and policy development for the cities and the urban regions of the transactional age. I am not the first to call for such a theory. My colleague Howell Baum, of the University of Maryland's Community Planning Program, saw this need many years ago (Baum, 1977).

It has been the premise here that today urban and regional planning practice throughout many of the world's industrial market economies is in a state of paradigm crisis. In essence, the crisis exists because old planning procedures and concepts of how the industrial city functions do not seem to apply today and for tomorrow. What is needed is a new planning paradigm (Hall, 1982: 298–303). March and Simon (1958) have made a distinction between substantive and procedural planning. Although such a distinction is well entrenched in planning theory (Hightower, 1969), the need that should be addressed is to explore and articulate a paradigm *for planning and policy practitioners* that is (1) both substantive *and* procedural and (2) contextual and derived form the postindustrial information society of the latter part of the twentieth century (Masuda, 1980). In concluding their retrospective view of the evolution of planning models and phases over the last 25 years or so, Galloway and Mahayni (1977) have stated that "at the operational level, the dynamics of planning action involve an interplay of procedural and substantive elements" (1977: 68).

For the last several years I have called for more transactional metropolitan knowledge generation and the utilization of that knowledge in the formulation of future policies (Corey, 1983, 1984b). This review suggests that, indeed, a great deal of substantive transactional urban research has occurred and is under way. However, I believe that there is a need now for more research and practice attention to be directed to *the procedural and planning-process aspects of transactional urbanization*. Such an emphasis would complement the growing substantive knowledge about the transactional metropolis. The interaction of transactional urban substance with transactional urban planning and policy processes is believed to be a means to the end of realizing a new planning paradigm—a "transactional metropolitan planning paradigm."

A fruitful next step toward the rigorous development (Teitz, 1981) of a transactional metropolitan planning paradigm might consist of applying a planning procedure and approach to the content of metropolitan-scale transactional economic, social, and physical functions. For a start, I would adopt the Program Planning Model noted above. PPM seems appropriate to steering transactional urbanization (Corey, 1985) because it is experimental, developmental, and it is both a social-behavioral and a rational-technical approach. Such procedural attributes seem well suited to empirical testing in transactional societies, cities, and organizations. PPM applies in uncertain, complex, turbulent environments (Emery, 1967) where innovation and new programs are needed to guide planning and policy actions. In addition to its use in the planning of transactional regional policies for Seoul's future (Corey, 1981), PPM already has been used to formulate strategies for neighborhoods (Corey, 1979) and to analyze urbanization patterns in Sri Lanka (Corey, 1984a).

I believe that the challenge now is for each of us to reflect on the current state of knowledge about the transactional metropolis, and to apply the best and most appropriate of our planning theory and methods to attain intentional and desired futures for our cities—both winter cities and otherwise—wherein transactional activities are present and determinant.

REFERENCES

ANDERSON, M. (1985) "Preface to transactional urban planning: employment dynamics in the San Francisco Bay Area." Unpublished master's dissertation, Institute for Urban Studies, University of Maryland, College Park.

BAUM, H. S. (1977) "Toward a post-industrial planning theory." Policy Sciences 8: 401-421.

BRUNN, S. D. and J. O. WHEELER [Eds.] (1980) The American Metropolitan System: Present and Future. New York: V. H. Winston.

BURKS, A. W. (1984) Japan: A Postindustrial Power. Boulder, CO: Westview.

CASTELLS, M. [Ed.] (1985) High Technology, Space, and Society. Beverly Hills, CA: Sage.

CLARK, D. (1985) Post-Industrial America: A Geographical Perspective. New York: Methuen.

CONZEN, M. P. (1983) "American cities in profound transition: the new city geography of the 1980s." Journal of Geography (May-June): 94-102.

COREY, K. E. (1979) Neighborhood Grantsmanship: An Approach for Grassroots Self-Reliance in the 1980s. Cincinnati, OH: Community Human and Resources Training, Inc.

COREY, K. E. (1980) "Transactional forces and the metropolis: towards a planning strategy for Seoul in the Year 2000," pp. 54-89 in Wo Kim (ed.) The Year 2000: Urban Growth and Perspectives for Seoul. Seoul: Korea Planners Association.

COREY, K. E. (1981) "The transactional metropolitan paradigm: an application to planning the metropolitan region of Seoul, Korea," pp. 39–49 in Proceedings of Applied Geography Conferences, (J. W. Frazier and B. J. Epstein, eds.). Binghamton, NY: Department of Geography, State University of New York.
COREY, K. E. (1982) "Transactional forces and the metropolis." Ekistics 297: 416–423.
COREY, K. E. (1983) "The transactional city: a call for research and policy attention." The Pennsylvania Geographer XXI, 3 and 4: 1–6.
COREY, K. E. (1984a) "Deconcentrated urbanization in Sri Lanka: a case of policy serendipity." Urban Studies Working Paper 2, College Park: Institute for Urban Studies, University of Maryland.
COREY, K. E. (1984b) "The transactional city: a call for research and policy attention." Geography Research Forum 7: 74–78.
COREY, K. E. (1985) "Qualitative planning methodology: an application in development planning research to South Korea and Sri Lanka." Urban Studies Working Paper 3, College Park: Institute for Urban Studies, University of Maryland.
DANIELS, P. (1982) Service Industries: Growth and Location. Cambridge: Cambridge University Press.
de BETTENCOURT, J. S., W. WIEWEL and R. MIER [Eds.] (1984) "High technology and economic development planning." Journal of the American Planning Association 50, 3: 262–296.
DELBECQ, A. L. (1978, March) "Relating need assessment to implementation strategies: an organizational perspective." Presented at the Second National Conference on Need Assessment in Health and Human Services, Louisville, Kentucky.
DWYER, D. J. (1985) "Urban growth, form and density: the transactional city and its counterparts in the third world," in Proceedings of the U.K.-Hong Kong Symposium on Environmental and Social Development, (P. Hills, ed.). Hong Kong: Centre of Urban Studies and Urban Planning, University of Hong Kong.
EMERY, F. E. (1967) "The next thirty years: concepts, methods, anticipations." Human Relations 20, 3: 199–237.
FUCHS, C. (1983) "Developers and users of office space: an examination of the location decision processes in the Washington, D.C. area, 1981–83." Master's dissertation, Department of Geography, University of Maryland, College Park.
GALE, D. E. (1984) Neighborhood Revitalization and the Postindustrial City: A Multinational Perspective. Lexington, MA: Lexington Books.
GALLOWAY, T. D. and R. G. MAHAYNI (1977) "Planning theory in retrospect: the process of paradigm change." Journal of the American Institute of Planners 43, 1: 62–71.
GAPPERT, G. and R. V. KNIGHT [Ed.] (1982) Cities in the 21st Century. Beverly Hills, CA: Sage.
GLASMEIER, A., P. HALL, and A. R. MARKUSEN (1984) "Metropolitan high technology industry growth in the mid 1970's: can everyone have a slice of the high-tech pie?" Berkeley Planning Journal 1, 1: 131–142.
GOTTMANN, J. (1955) Virginia at Mid-Century. New York: Henry Holt.
GOTTMANN, J. (1961) Megalopolis: The Urbanized Northeastern Seaboard of the United States. New York: The Twentieth Century Fund.
GOTTMANN, J. (1970) "Urban centrality and the interweaving of quarternary activities." Ekistics 29, 174: 322–331.
GOTTMANN, J. (1971) "Pour une geographie des centres transactionnels." Bulletin De L'association Geographes Francais (Paris), January-February (No. 385–386): 41–49.

GOTTMANN, J. (1972) "The city is a crossroads." Ekistics 34, 204: 308-309.
GOTTMANN, J. (1975) "The evolution of urban centrality." Ekistics 39, 233: 220-228.
GOTTMANN, J. (1978) "Urbanization and employment: towards a general theory." Town Planning Review 49, 3: 393-401.
GOTTMANN, J. [Ed.] (1979) "Offices and urban growth." Ekistics 46, 274: 3-66.
GOTTMANN, J. (1983a) "Capital Cities," Ekistics 50, 229: 88-93.
GOTTMANN, J. (1983b) The Coming of the Transactional City. College Park: Institute for Urban Studies, University of Maryland.
GUENTHER, R. (1985) "Conference centers catch on as some firms shun hotels." Wall Street Journal (March 6): 37.
HALL, P. (1981) "Issues for the eighties." The Planner 67, 1: 4-5.
HALL, P. (1982) Urban and Regional Planning. Harmondsworth, England: Penguin.
HALL, P. (1984) "Have cities a future?" Futures 16, 4: 344-350.
HALL, P. (1985) "The decline of the cities: a problem with its roots in the distant past." Town and Country 54, 2: 40-43.
HALL, P. and A. MARKUSEN [Eds.] (1985) Silicon Landscapes. Boston: Allen & Unwin.
HARPER, R. A. (1982a) "Metropolitan areas as transactional centers," pp. 87-109 in C. M. Christian and R. A. Harper (eds.) Modern Metropolitan Systems. Columbus, OH: Merrill.
HARPER, R. A. (1982) "The transactional society: the need for a different dimension to geographic research." Presented at the Fifth Annual Applied Geography Conference, College Park, Maryland.
HIGHTOWER, H. C. (1969) "Planning theory in contemporary professional education." Journal of the American Institute of Planners 35: 326-329.
HOFHEINZ, R., Jr., and K. CALDER (1982) The East Asia Edge. New York: Basic Books.
HOWLAND, M. (1984, October) "Why are plant closing rates so high in the sun-belt?" Presented at the Association of Collegiate Schools of Planning Conference, New York City.
JACOBS, J. (1984) Cities and the Wealth of Nations: Principles of Economic Life. New York: Random House.
KELLERMAN, A. (1984) "Telecommunications and the geography of metropolitan areas." Progress in Human Geography 8, 2: 222-246.
KELLERMAN, A. (1985) "The evolution of service economies: a geographical perspective." The Professional Geographer 37, 2: 133-143.
KELLERMAN, A. (1986) "Characteristics and trends in the Israeli service economy." The Service Industries Journal 6, 2: 205-226.
KOENIG, R., Jr. (1984) "More cities rush to host trade shows." Wall Street Journal (May 24): 35.
LANGDALE, J. V. (1982) "Telecommunications in Sydney: towards an information economy," pp. 72-94 in R.V. Cardew et al. (eds.) Why Cities Change: Urban Development and Economic Change in Sydney. Sydney: Allen & Unwin.
LANGDALE, J. V. (1984a) "Computerization in Singapore and Australia." The Information Society 3, 2: 131-153.
LANGDALE, J. V. (1984b) Information Services in Australia and Singapore. Kuala Lumpur and Canberra: ASEAN-Australia Joint Research Project.
LANGDALE, J. V. (1985) "Electronic fund transfer and the internationalization of the banking and finance industry." Geoforum 16, 1; 1-13.

MARCH, J. G. and H. A. SIMON (1958) Organizations. New York: John Wiley.
MASUDA, U. (1980) The Information Society as Post-Industrial Society. Tokyo: Institute for the Information Society.
McKOY, J. H. (1984) "Proposed research project." Manuscript, University of Maryland, Institute for Urban Studies, College Park, Maryland.
NAGASHIMA, C. (1981) "The Tokaido megalopolis." Ekistics 48, 289: 280–301.
NAISBITT, J. (1982) Megatrends: Ten New Directions Transforming Our Lives. New York: Warner Books.
NOYELLE, T. J. and T. M. STANBACK, Jr., (1983) The Economic Transformation of American Cities. Totowa, NJ: Rowman & Allanheld.
PATTEN, J. [Ed.] (1983) The Expanding City: Essays In Honour of Professor Jean Gottmann. London: Academic.
PERLOFF, H. S. (1980) Planning the Post-Industrial City. Washington, DC: Planners Press, American Planning Association.
ROGERS, E. M. (1985) The High Technology of Silicon Valley. College Park: Institute for Urban Studies, University of Maryland.
SMOUT, M.A.H. (1983) "Geography and the post-industrial city." Journal of the University of Durban-Westville 4, 2: 191–198.
STAFFORD, H. and J. REES (1984) Technology, Innovation, and Regional Economic Development. Washington, DC: Office of Technology Assessment (OTA-STI-238).
STANBACK, T. M. et al. (1981) Services—The New Economy. Totowa, NJ: Allanheld, Osmun.
TEITZ, M. B. (1981) "Review of Planning the Post-Industrial City." Journal of the American Planning Association 47, 3: 353–355.
VAN DE VEN, A. H. and R. KOENIG, Jr. (1976) "A process model for program planning and evaluation." Journal of Economics and Business 28, 3: 161–170.
YOUNG, K. and L. MILLS (1983) Managing the Post-Industrial City. London: Heinemann Educational Books.

9

The Great Lakes Megalopolis

HAROLD M. MAYER

□THE GREAT LAKES form a major part of the axis of what has long been identified as the "core region" of North America (Patterson, 1984: 186; Ullman, 1957; White et al., 1985). This region extends from the northeastern seaboard of the United States and the St. Lawrence valley of Canada to and beyond the western Great Lakes (Watson, 1982). Within this region two dominant subregions can be identified. One is the lineal conurbation extending from southern New Hampshire to northern Virginia in a northeast-southwest direction along the "fall line," the zone of contact between the hard-rock piedmont and the Atlantic coastal plain. Along this line, at the head of navigation of the rivers and estuaries, where waterfalls and rapids furnished power during the early days of settlement, urban centers developed. Their early growth was stimulated by the combination of navigation access and water power along the lineal axis (Van Cleef, 1957). This spatial pattern of urban settlement, common in other regions, was identified by geographers as one of the three basic interurban settlement forms. Jean Gottmann called the northeastern seaboard conurbation "Megalopolis" (1961). The other lineal conurbation within the core region of the continent has as its axis the St. Lawrence-Great Lakes system extending from the head of the St. Lawrence estuary at Quebec city, to beyond the western Great Lakes, with branches and extensions of high-density urban settlement along the Hudson-Mohawk corridor in New York State, and from east of Pittsburgh to Lake Erie (Doxiadis, 1966; Leman Group Inc., 1976). This "Great Lakes megalopolis" is separated from the northeastern megalopolis by a short gap across the Appalachians in central Pennsylvania. The two megalopoli together form the core area of the continent, containing the preponderant population of both the United States and Canada, as well as the densest concentration of industrial and commercial activity and the most intensively used transportation routes (Wade, 1969; Yates, 1975).

Along the shores of the Great Lakes and the St. Lawrence River and in the immediate hinterland are 8 of the 50 most populous metropolitan areas of the United States and 11 of the 18 most populous of Canada.[1]

Most of the industrial, commercial, and agricultural activity of the Great Lakes region was attracted originally by the availability of water transportation. Once established, the centers in turn attracted overland transportation: at first the railroads, later the major highways, and more recently airlines, thus reinforcing the prominence of the earlier settled centers, some of which grew into metropolises of world importance. In the United States these include Chicago, the primate city of the continental interior (Mayer and Wade, 1969), as well as Rochester, Buffalo, Pittsburgh, Cleveland, Toledo, and Detroit. In Canada the two largest metropolitan areas, Toronto and Montreal, are both situated along the Great Lakes-St. Lawrence axis; the national capital, Ottawa, is a short distance from the axis, as is Winnipeg, the major industrial and transportation center associated with the Canadian lakehead area (Whebell, 1969).

Although facing losses of relative, and in some instances absolute, population and industrial activity in recent decades (and especially in the 1970s and early 1980s), the cities and metropolitan areas of the Great Lakes megalopolis still, along with the North Atlantic seaboard of the United States, form the dominant nodal region of North America (Federal Reserve Bank of Chicago, 1985). In spite of the decline of the traditional "smokestack" industries that dominated the economy of the region for more than a century, and the recent growth of the service industries, the prospect is that the Great Lakes region will continue to be a major focus of population and employment in the forseeable future.[2]

As previously, transportation continues to play a dominant role in the economic base of the region, although recently the modes of transportation have changed significantly in relative importance. The significance of the Great Lakes-St. Lawrence waterway system is in some respects different than was anticipated when the St. Lawrence Seaway was planned.

Within the Great Lakes region there is a complementary of the resource base as well as of occupance patterns and urban development between the northern Lakes area, which is dominated by natural resources—iron and copper ore and forests—and the southern

Lakes region, where urban activities predominate. The division between the two complementary areas follows approximately a line just north of peninsular Ontario and extending westward at the southern end of Lake Huron, across the southern peninsula of Michigan through Bay City and Ludington, thence through Green Bay and central Wisconsin (Mayer, 1969).

This complementary stimulated the development, and in turn was reinforced by, a system of water transportation that has tied together the two complementary portions of the Great Lakes region. The north-south orientation of internal Great Lakes commerce for over a century was manifest in a predominance of bulk commodity movements: iron, and, to a lesser extent, copper, ores, from the northern lakes area to the industrial cities on the shores of lakes Erie and Michigan and in their hinterlands; cereal grains from the agricultural hinterlands to the west of Lake Superior (wheat) and Lake Michigan (principally corn) to the cities of the Great Lakes megalopolis as well as to overseas destinations; coal from Appalachia and the interior coalfields to the iron and steel and electric generating plants around and near the lakes, and forest products from, the northern lakes area to the southern consuming areas as well as to the west through the Great Lakes ports. All of these internal Great Lakes movements are still important, although the volume has not increased for many years and waterborne commerce on the Great Lakes has not shared in the overall increase of goods movements within the United States and Canada. But the significance of bulk commodity transportation on the Great Lakes is much greater than the trends in tonnage movements would indicate. This is because, even in decline, many of the industries that depend upon such movement continue to be essential to the national economies of the two nations bordering the lakes. Furthermore, not all of the "high-tech" industries that in part have replaced the "smokestack" industries in the region and elsewhere are free from dependence upon the natural resources of the Great Lakes region. The increasing use of computers, for example, is resulting in an increasing demand for paper, the production of which is a major industry of the region. Wisconsin is the leading producer of paper in the United States (Wisconsin Strategic Development Commission, 1985).

The Great Lakes have been connected with overseas ports by all-water routes since the early nineteenth century: via the Erie Canal-Hudson River route with transhipment between lake and canal vessels, and through a system of Canadian canals circumventing

Niagara Falls and the St. Lawrence rapids—by direct Great Lakes over-seas movement of small oceangoing vessels, as well as by transhipment from lake to smaller "canaller" vessels and thence again to oceangoing ships in the lower St. Lawrence—since the 1820s (Mayer, 1954). The enlarged St. Lawrence Seaway, opened in 1959, did not thus constitute a new all-water route, but rather the improvement of a route that existed for many decades. The present waterway system, as a whole, is intimately related to the economic base of the entire Great Lakes region, and, more specifically, to the Great Lakes ports and their respective metropolitan areas and hinterlands, constituting the Great Lakes Megalopolis.

The Great Lakes region continues to grow, although at a slower rate than that of either the United States or Canada. Between 1950 and 1980 the proportion of the United States population in the region has declined from slightly over 20% to about 18% (Federal Reserve Bank of Chicago, 1985). In the period immediately following World War II, the rate of population increase in the region was greater than that of the nation as a whole, but since the 1950s the rate has averaged half that of the nation. The region's largest metropolitan areas all suffered a decline in their rate of growth, and 5 of the 11 largest were subjected to an absolute population decrease. In Canada the two provinces bordering the Great Lakes-St. Lawrence waterway—Ontario and Quebec—had a decline of from 64% to 62% of that nation's population between 1971 and 1981, whereas both Toronto and Montreal, in spite of a 7% increase of the former, fell far behind the rate of growth of the major western Canadian cities: Calgary, Edmonton, and Vancouver.[3]

Migration tends to follow economic opportunity and to a somewhat lesser degree is conditioned by amenity considerations, especially climate, as work tends to occupy a lesser proportion of a person's lifetime (Ullman, 1954). Population shifts produce, and in turn result from, shifts in market location: the demand for goods and services. The Great Lakes region witnessed a decline in its economic base since the 1960s, partly as a result of interregional population shifts and, in turn, partly as the result of changes in the "mix" of employment opportunities. A decline has occurred in traditional production of durable goods and there is a greater emphasis on high-tech industries and service activities, both of which are less dependent upon transportation of large volumes of goods and are more subject to the amenity characteristics of locations.

Between 1970 and 1980 the Great Lakes region's nonagricultural employment declined from 20.6% to 18.6% of that of the United States as a whole (U.S. Bureau of the Census, 1985). In spite of this decline the 1980 census revealed that the region continued to lead the nation in nonagricultural employment. How long this will continue constitutes a challenge to the region in view of the trends just outlined.

Prospects for future industrial expansion in the Great Lakes region are not encouraging, although there are indications that the decline may have bottomed out with the economic recovery that began in late 1982. The motor vehicle manufacturing industry, one of the region's largest industrial employers and also the prime customer of the region's iron and steel industry, has led the recovery. Within the five states of the Chicago Federal Reserve District—Illinois, Indiana, Michigan, Wisconsin, and Iowa—is located about 58% of the nation's employment in that industry; Michigan alone accounts for 44% (Federal Reserve Bank of Chicago, 1985). In addition, there are large motor vehicle manufacturing plants in Ohio and western Pennsylvania, within the Great Lakes region. In the industry as a whole, employment reached a peak of over 1 million in late 1978, declined to under 700,000 in 1982, recovered to 880,000 in late 1984, but was still 16% below the earlier peak (Federal Reserve Bank of Chicago, 1985). Iron and steel production in late 1984, in turn, was more than double that of two years earlier, in spite of the fact that many of the older and more obsolescent plants, formerly employing many thousands, had closed or were in partial operation. On the other hand, another major customer of the iron and steel industry—farm equipment manufacturing—continued to decline substantially, with a decrease of over 12% in value of tractor production and more than 60% in combines in late 1984 as compared with two years earlier. (Federal Reserve Bank of Chicago, 1985). This reflects the more acute and precarious position of many Midwest farmers, a significant proportion of whom were on the verge of bankruptcy.

The relative long-term decline in the economic base—and hence employment and, ultimately, of population of the Midwest, which includes most of the large metropolitan areas of the Great Lakes Megalopolis—is shown by the proportion of the United States population in the Midwest (designated as the North Central region until 1984) as released by the U.S. Census in December 1984. It shows that the population of the midwestern states was virtually stable since the 1980 census. The relative decline, however, has continued. In 1930 the region contained 31.3% of the nation's population; in 1960 it was

28.8%, and in 1980 it was down to 26.0%. By mid-1984 the proportion was estimated to have declined to 25%. In spite of this relative decline, the Midwest still leads the Northeast and the West, which are estimated to have had in mid-1984 about 21.1% and 19.8%, respectively, of the nation's population. Only the South, with 34.1%, has a higher proportion of the national population than the Midwest. On the other hand, the South and West together are estimated to have accounted for 91% of the national population increase between 1980 and 1984.

The most important problem facing the cities and metropolitan areas of the Great Lakes region is that of adjusting to the slower rates of population and economic growth characterizing the recent past and which are in prospect for the future. In the traditional industries that heretofore have constituted major elements of the region's economic base, prospects for increased employment are far from encouraging. The iron and steel industry is faced with decreasing demand, not only because of competition from imported steel originating in countries where labor is much less costly, but also from foreign competition in the transportation equipment manufacturing industries. Motor vehicle manufacturers must compete with foreign producers, who have increased their market share to as much as one-quarter of the domestic United States and Canadian markets. Railroad and urban transit equipment manufacturers, who also constitute important components of the Great Lakes region's economy, also face foreign competition: new transit cars for Cleveland came from Italy, bus bodies for Milwaukee were manufactured in Hungary. Farm equipment manufacturers, also formerly important consumers of steel produced in the Great Lakes region, face competition from imports from Canada, Japan, and the Soviet Union, among other countries. Demand for railroad rails is slight, due to the availability of second-hand rails resulting from reduction in the size of the railroad networks of both the United States and Canada, in part resulting from deregulation. Employment in iron and steel manufacturing cannot be expected to increase as rapidly as any possible increase in production because of technological efficiencies resulting from such innovations as the substitution of basic oxygen processing instead of open-hearth furnaces, continuous casting, replacement of much iron ore by scrap processed in electric furnaces, and, of course, importation of steel. It is ironic that the steel industry of the Great Lakes region, whose advocacy of the St. Lawrence Seaway for exploitation of the ores of the Quebec-Labrador region in the 1950s made the seaway politically possible, now faces the competition of imported steel entering the region in part because of low-cost transportation through the seaway.

One of the economic problems of the metropolitan areas in the Great Lakes region is caused by the trend toward larger organizational units resulting from mergers. Not only do the resultant economies of scale result in many instances of decreased employment, but the mergers commonly result in the relocation of administrative headquarters as well as production employment away from the region to other locations, commonly in the Sun Belt but also abroad. Among the examples are American Motors, affiliated with a French manufacturer; Kimberly Clark, a major forest products firm that moved its headquarters from Wisconsin to Texas; Allen-Bradley, which merged with another firm and reduced its Wisconsin employment; and several large retailers, including Marshall Field, important in the Great Lakes region but now a subsidiary of a British conglomorate.

On the other hand, in several of the larger Great Lakes cities the demand for office space has increased rapidly (Urban Investment and Development Company, 1984). International, national, and regional headquarters of many commercial and industrial firms have burgeoned. Chicago has maintained its position as the primate city and metropolitan area of the region, with a virtually new office skyline both in the Loop district and along North Michigan Avenue on the city's near north side. At the same time there has been rapid expansion of office building clusters in several of the outlying areas near the city, especially in DuPage County to the west and in the vicinity of O'Hare International Airport to the northwest. O'Hare has maintained its position as the world's busiest air carrier airport, with over 30,000 employees within the airport boundary and many thousands more in the immediate vicinity. With the extension of rapid transit to O'Hare in 1984 several large clusters of office buildings, hotels, and convention facilities have arisen, their magnitude rivaling the central business districts of many midsized cities (Chicago Association of Commerce and Industry, 1984). In the Detroit area, large complexes have developed at Southfield and other northern suburbs, as well as the Renaissance Center adjacent to the city's downtown area. Toronto, Cleveland, Buffalo, and Milwaukee, among other cities, have been and are building substantial new downtown projects. In the immediate Great Lakes hinterland, the twin cities of Minneapolis and St. Paul are witnessing expansion of their office, hotel, convention, and sports facilities downtown, with concurrent developments in several of the suburban areas.

Thus, there is considerable development, both centrally and peripherally, in the major metropolitan areas of the Great Lakes region. Much of the development is different in character from that which until recently characterized the urban portions of the Great Lakes region, with accommodations for service activities and amenities—both of which produce substantial employment directly—replacing major portions of the heavy manufacturing infrastructure, which has been declining rapidly in recent years. On the other hand, although the newer high-tech industries are not growing in the Great Lakes region as rapidly as in portions of the Sun Belt, development of such industries is far from lacking. There are many continuing, new, and expanding establishments in the service, data processing, aerospace, and other categories. Some of the earlier research and development facilities as well as high-tech manufacturing establishments remain. Among the latter are the General Electric plants in Erie, Pennsylvania, and Nela Park in Cleveland, as well as the General Motors and other motor vehicle research laboratories in Michigan. In recent decades notable new research and development and high-tech manufacturing establishments in the Great Lakes region include the Argonne and Fermilab basic science facilities in the Chicago area, the large facilities of Xerox near Rochester, New York, and General Electric's medical electronics activities near Milwaukee, among others.

Like other urban complexes throughout North America, the metropolitan areas of the Great Lakes region are expanding spatially. As elsewhere, urban densities in the peripheral areas and, in many instances the "inner city" renewed areas as well, are subject to lower population densities than those of the earlier-developed areas. The result is that the demand for land continues to increase, especially along and beyond the urban peripheries, at a much greater rate than the growth—where it occurs—of population. Suburbanization continues, but most rapidly along the corridors of transportation, and especially in the vicinities of the new express highways, including the interstates (Highway Research Board, 1965).

The peripheral expansion of urbanization along the transportation corridors, and especially along the express highway axes, is gradually increasing the overlap between nearby metropolitan areas. There is substantial cross-commuting in the multinucleated urban complex of northeastern Ohio (Mayer and Corsi, 1976), the Detroit-

Toledo area, and along the Chicago-Milwaukee corridor (Cutler, 1965). Intermetropolitan coalescence is increasingly prominent, offering the opportunities in many instances for choosing employment in more than one metropolitan area from a given residence location. It is not unusual for families with more than one employed member to choose a residential location in an area convenient to employment, recreational, and cultural opportunities in more than a single metropolitan area.

The peripheral population increases have their reciprocal in decreases in many of the central cities of the Great Lakes region, as elsewhere. In spite of "gentrification" and the consequent return of numerous middle-income families and individuals to the central portions of many of the Great Lakes cities, the general decrease of populations in the inner city areas continues. It is commonly accompanied by an increasing separation of socioeconomic groups, as well as racial segregation, with, in many instances, intensification of the problems of ethnic and racial segregation. Great Lakes cities, of course, are not alone in these problems.

In summary, the Great Lakes megalopolis is a region of changing economic characteristics, many of which are not unique to the region. The urban areas of the region, in general, are declining in relative, though not necessarily in absolute, population growth. They face serious problems of adjustment to the general shift from "heavy" manufacturing to service activities, from capital-intensive activities to those involving high tech, in which competition from Sun Belt areas and from abroad constitute challenges. The Great Lakes themselves, although continuing to be important for some types of transportation, are somewhat less significant in terms of the transportation needs of the most rapidly growing types of industrial and commercial activities. On the other hand, the Great Lakes offer unlimited supplies of fresh water, an advantage that many portions of the Sun Belt lack; this may, in the long run, compensate to some degree the disadvantage of the severe winter climate of the Great Lakes region in attracting some types of industrial and commercial activities. Although faced with serious external competition and internal problems, the Great Lakes megalopolis is still a dominant locus of population and economic activity.

NOTES

1. Compiled from various publications of the U.S. Bureau of the Census and Statistics Canada.

2. The growth of service industries in the major Great Lakes metropolitan areas is compensating to some degree for the decline in "heavy" manufacturing.
3. Information from Statistics Canada.
4. Information from a press release of the U.S. Bureau of the Census.

REFERENCES

Chicago Association of Commerce and Industry (1984) "O'Hare/Woodfield." Commerce 81, 8: 33ff.

CUTLER, I. (1965) The Chicago-Milwaukee Corridor: A Geographic Study of Intermetropolitan Coalescence. Evanston, IL: Northwestern University. Department of Geography.

DOXIADIS, C. A. (1966) Emergence and Growth of an Urban Region: The Developing Detroit Area, Vol I: Analysis. Detroit: Detroit Edison.

Federal Reserve Bank of Chicago (1985) Midwest Update. No. 34 (March 1).

GOTTMAN, J. (1961) Megalopolis: The Urbanized Northeastern Seaboard of the United States. New York: The Twentieth Century Fund.

Highway Research Board (1965) Indirect and Sociological Effects of Highway Location and Improvement. Highway Research Record No. 75, Washington, D.C.

Leman Group Inc. (1976) Great Lakes Megalopolis: From Civilization to Ecuminization. Ottawa: Supply and Services Canada.

MAYER, H. M. (1954) "Great Lakes overseas: an expanding trade route." Economic Geography 30, 2: 117–143.

——— (1969) "The Great Lakes: the tie that binds." American Institute of Architects Journal (June): 50–58.

——— and T. CORSI (1976 "The northeastern Ohio urban complex," pp. 109–179 in J. S. Adams (ed.) Contemporary Metropolitan America. Cambridge, MA: Ballinger.

MAYER, H. M. and R. C. WADE (1969) Chicago: Growth of a Metropolis. Chicago: University of Chicago Press.

PATERSON, J. H. (1984) North America. New York: Oxford University Press.

URRMANN, E. L. (1954) "Amenities as a factor in regional growth." Geographical Review 44, 1: 19–32.

——— (1957) American Commodity Flow. Seattle: University of Washington Press.

Urban Investment and Development Company (1984) Downtown Office Construction in Major U.S. Cities. Chicago: Author.

U.S. Bureau of the Census (1985) Statistical Abstract of the United States, 1985. Washington, DC: Government Printing Office.

WADE, M. [ed.] (1969) The International Megalopolis. Toronto: University of Toronto Press.

WATSON, J. W. (1982) The United States. New York: Longman.

WHEBELL, C.F.J. (1969) "Corridors: a theory of urban systems." Annals of the Association of American Geographers 59, 1: 1–26.

WHITE, C. L., E. J. FOSCUE, and T. L. McKnight (1985) Regional Geography of Anglo-America. Englewood Cliffs, NJ: Prentice-Hall.

Wisconsin Strategic Development Commission (1985) Phase I, The Mark of Progress. Madison: Author.

Van CLEEF, E. (1937) Trade Centers and Trade Routes. New York: Appleton-Century-Crofts.

YEATES, M. (1975) Main Street: Windsor to Quebec City. Ottawa: Macmillan of Canada Ltd.

10

The Nation's Most Livable City: Pittsburgh's Transformation

BRIAN J.L. BERRY, SUSAN W. SANDERSON, SHELBY STEWMAN, and JOEL TARR

□THE REVIVAL and restructuring of the Pittsburgh region did not occur by chance. Both internal and external factors have contributed to the renaissance that is now taking place. Essential foundations were laid internally by a public-private partnership, the most dramatic product of which has been the physical reconstruction of downtown. External factors, especially the sharp recession of the early 1980s, administered a final coup de grace to an already declining steel industry, enabling a new economic base to become ascendant. From these shifts has emerged a modern producer-service economy, proud of the surprise that Rand McNally's rating as the nation's "most livable" city elicits. In this chapter we describe Pittsburgh's public-private partnership, document the transformation of the economic base, and examine some of the social and geographic consequences of this transformation.

THE PITTSBURGH TRADITION OF PUBLIC-PRIVATE PARTNERSHIP

Beginning in the immediate post-World War II period the city faced population losses and economic decline. Outstanding leaders on each side perceived an environment of opportunity for major urban redevelopment. Trust between the leadership resulted in a wide-ranging series of developments known as Renaissance I. Changing times brought different leaders with different values and agendas, and the original partnership dissolved, only to be renewed again under somewhat different conditions. Pittsburgh is now in the middle of a second renaissance brought about by a revised version of the old public-private partnership. Simultaneously, the region is undergoing major structural change. The challenge for the tradition of public-private partnership is its ability to respond creatively to the new challenges facing the city and the region.

Although these partnership arrangements have been most significant in Renaissance I and Renaissance II, they were also important in the earlier part of the century. In the pre-World War I years, for instance, private-sector groups such as the Chamber of Commerce and the Civic Club led the way toward rationalizing and modernizing governmental institutions and improving environmental quality. There were some successes, particularly in the direction of centralization of governmental authority and uniformity in education and taxation, but there were many failures in improving the basic quality of urban life. By World War II, however, a number of knowledgeable Pittsburghers were convinced that only sustained intervention by both public and private groups under the direction of new leadership and utilizing new institutions could prevent the city from experiencing severe corporate and population losses in the postwar period.

RENAISSANCE I

The shared perception of impending crisis galvanized key figures in the city to form a new organization, the Allegheny Conference on Community Development (ACCD), and to persuade Pittsburgh's single most powerful businessman, Richard K. Mellon, to take the leadership in urban redevelopment. The ACCD was primarily a business group, composed of the chief executive officers of the city's major corporations. Early in its deliberations the conference adopted a critical rule that no delegation of voting power was permitted. That is, unless members were at the meeting where the votes were taken they would lose their vote. Working with its planning adjuncts, the Pennsylvania Economy League and the Pittsburgh Regional Planning Association, the ACCD made a number of critical decisions regarding the development of the city and the region and generated the blueprints to provide the necessary guidelines.

Implementation of massive renewal plans, however, could only take place with the cooperation of the public sector. Here the key figure was David L. Lawrence, the head of the Democratic organization and mayor of the city from 1945 to 1961. When Lawrence ran for mayor, he did so on a platform that included a statement that he would aid and support the ACCD plan for the improvement of

Pittsburgh. Upon his election he proceeded to appoint key figures from the ACCD to various appointive posts, laying the groundwork for one of the most significant instances of public-private partnership in the history of American cities.

The major accomplishments of Renaissance I were environmental improvements and renewal of the central business district. These involved governmental action by city, county, state, and federal authorities as well as by the private sector. The normal pattern of development in the early years of Renaissance I was for plans to be developed by the ACCD and its nonprofit planning adjuncts in consultation with public representatives. These plans would then be transmitted to the city for implementation or to one of the several important authorities created by the city such as the Urban Reconstruction Authority (URA) or the Parking Authority. These appointed authorities had boards of directors staffed from both the public and private sectors. Using the power of eminent domain, the URA condemned areas throughout the Golden Triangle, the city's central business district, and the city provided the infrastructure and private investment provided for construction.

The clear precondition for investment to spur redevelopment was environmental improvement. Most important were smoke and flood control. Smoke control essentially involved implementing a law passed in 1941 that required Pittsburghers to substitute clean fuel for dirty but cheap bituminous coal. With municipal and private groups working together toward effective implementation, the law was enforced beginning in the winter of 1948. Supplies of cheap natural gas combined with effective public policy to free Pittsburgh skies of the blanket of heavy smoke that had filled them for many years. Control of the floods that periodically inundated the city was also a necessary precondition for securing new investment in the downtown. Here, action by the federal government was spurred by the public-private forces to complete a string of eight flood-control dams begun in the pre-World War II period.

In the Golden Triangle the most consequential developments involved clearing land of old and decayed commercial and industrial structures and replacing them with parks, modern high rise office structures, and a civic arena. The first and most significant step was to clear the land in the triangle formed at the confluence where the Allegheny and Monongahela rivers form the Ohio River. This was the city's original location, at Fort Duquesne. The cleared area became Point State Park. This action required coordination between the Republican-dominated state government and the city administration through Point State Park Commission. Significantly, the commission was headed by Richard K. Mellon's right-hand man, Arthur Van Buskirk. Simultaneously, the land abutting the park was developed as an office complex called Gateway Center with funds provided by the Equitable Insurance Company. Here the newly created Urban Redevelopment Authority (URA), led by a mix of public and private sector figures appointed by the mayor, was instrumental in land acquisition and clearance. In this case, a precedent was set in the use of powers of eminent domain for a commercial purpose.

Further to the east in the middle of the "Golden Triangle" Richard K. Mellon proposed the building of an urban park surrounded by office towers and hotels. With the utilization of funds supplied through one of the Mellon foundations and the powers of the URA, land was purchased, cleared, and a park with an underground garage constructed. Abutting the park at opposite corners, Mellon National Bank and ALCOA constructed 40-story central office headquarters buildings. Still further to the east, in an area occupied by the city's worst housing slum, the URA again used its powers to occupy and clear land—this time for the purpose of building a civic arena with a movable roof to house the city's various musical and sports entertainments. In this case, however, unlike the earlier cases of rebuilding, land acquisition involved the displacement of thousands of poor blacks and whites. The necessity for civic betterment was offered as a justification and it was claimed that most of the displaced residents were satisfactorily relocated, but the episode remains as a negative incident in Pittsburgh's renewal.

In addition to these environmental and redevelopment activities, actions were also taken in a number of other areas. Most significant on the private-sector side were the creation of a nonprofit association to attract new industry to the area (Regional Industrial Development Association), as well as the establishment of the first community-supported educational television station. On the public side, the establishment of the Allegheny County Sanitary Authority to improve the sewage collection and treatment, the expansion of the airport, and the creation of Port Authority Transit for Allegheny County were also important. No doubt, however, the environmental and bricks-and-mortar accomplishments stand out as the most significant accomplishments of Renaissance I.

THE INTERLUDE

By the 1960s, conditions had shifted greatly from the inception of Renaissance I. For one, leadership on both the public and private sides had changed, with the death of Mellon and with Lawrence moving on to become governor of the Commonwealth. While the redevelopment of the Golden Triangle had continued and white-collar office employment had grown, office occupancy rates had dipped as new office construction outdistanced employment. In addition, the city had suffered sharp population losses since 1950. Neighborhoods felt neglected because of the downtown focus, and resentment persisted over the population dislocations that occurred because of the projects such as the Civic Arena renewal, expansion of the Jones & Laughlin steel facility, the East Liberty Renewal, and the proposed East Street Expressway. At the same time, the city bureaucracy, under control of the Democratic organization, had grown considerably. Because of expenditures on city hall staff and on redevelopment, city taxes had risen. In the context of new thinking and values regarding community and democratic participation, many Pittsburghers began to question the breadth of the accomplishments of the renaissance and also the equity of the public partnership.

Development did not come to a halt in the interlude, nor did public-private partnerships come to an end. Rather, while the Renaissance I partnerships were being dissolved, a new type of partnership was being shaped: social partnerships between the public and private sectors. Two specific areas of change emerged: new links with the black community, and the reformulation of "urban renewal" in the area of neighborhood housing.

SOCIAL PARTNERSHIPS: THE BLACK COMMUNITY AND NEIGHBORHOODS

Important new public-private partnerships were developed in the 1960s to help fill the need for educational and job opportunities for the black population. The important groups here were the Negro Education Emergency Drive (NEED), the Program to Aid Citizen Enterprise (PACE), the National Alliance of Businessmen (NAB), and the Minority Entrepreneur Loan Program (MELP). Each program involved collaboration between members of the corporate and black communities. In terms of economic development, MELP is the most pertinent. It was organized in 1968 by the Economic Development

Committee of the ACCD to help minorities own and manage their own business. The program provided seed money in the form of a subordinated loan to enable a minority businessperson to obtain a loan from the Small Business Administration (which generally would lend up to 80% or 90% of the amount needed). Between 1968 and 1976, MELP made over $15 million in loans and loan commitments to 650 minority entrepreneurs. The failure rate was about the same as the national average for all new businesses. At its high point, there were more than 200 additional minority-owned enterprises in the Pittsburgh region as a result of MELP.

The second form of social partnership that developed during the interlude pertained to neighborhoods. In July 1968 the Central North Side Improvement Fund—soon to be renamed Neighborhood Housing Services (NHS)—was incorporated, a result of two years of work by neighborhood citizens' groups, the Mayor's Office, and financial institutions. NHS can be thought of in two ways: as a citizen-based effort to improve the neighborhood and as a revolving loan fund to improve housing in the area. It is the neighborhood aspect that distinguishes this program from most housing-related programs and clearly places it in the social development category.

The central north side neighborhood was clearly in serious trouble: its 3900 housing units were aging and in disrepair, the neighborhood was clearly declining, and it had been declared an urban renewal clearance area. Investment mentality in the area was pessimistic. The principal significance of NHS in Pittsburgh was not its dollar value (less than $1 million) but in the partnership that evolved and the associated individual, organizational, and financial commitments by all three of the principals (citizens' groups, local government, and financial institutions). Citizens were the heart of this program. They pressured the city to recognize and then raise its level of commitment to the neighborhood. They were the primary force in enabling NHS to obtain private funding for a revolving loan fund for high-risk residents; in persuading financial institutions to fund the administrative costs of the revolving loan fund and to provide technical and data processing services on the nonbankable loan accounts; in obtaining the community's acceptance of the code enforcement program; and in the continuing internal operation of NHS via its board of directors. What was distinctive was that citizens were partners with both local government and financial institutions.

PETER FLAHERTY AND THE POLITICS OF INDEPENDENCE

An important symbol of a new attitude toward the renaissance was, however, the election of a maverick Democratic city councilman, Peter Flaherty, as mayor of Pittsburgh in 1969 on a platform that he was "nobody's boy." He would work with neither the business establishment as represented by the ACCD, nor with the Democratic organization. Flaherty's election meant the temporary suspension of the public-private partnership. As mayor, he had a new agenda: reduce the size of the city work force, freeze taxes, and reorient spending from the downtown to the neighborhoods. Flaherty replaced the directors of all departments including those most directly involved in planning and development, the Planning Department, and the URA. On substantive development matters, he blocked the completion of two of the conference's favorite transportation projects—construction of a rapid transit system using a futuristic people-moving system called "Skybus" and the building of a super highway in the East Street Valley.

There were other elements in the Flaherty program that have proven to be important in the long run, however—particularly the restoration of a balanced budget, employment reorganization, and cutbacks in the city payroll. He instituted funding of the pension program in order to maintain a high bond rating for the city, and reduced permanent full-time employment by 27% over the seven years of his mayoralty. Reorganizations with an efficiency emphasis took place across the work force.

In regard to the neighborhoods, Flaherty redirected federal aid money from a primarily downtown orientation to a 50/50 split with the local areas. He instituted a home improvement loan program from Community Development Block Grants and constructed skating rinks and swimming pools in the city's parks. Also, downtown development did not halt and included the construction of several high-rise structures; the pace, however, was much less rapid than in the 1950s and 1960s. Given the fact that office occupancy rates had continually dropped since the late 1960s, this should be no surprise. More significant, at the beginning of Flaherty's second term as mayor, an agreement was reached between him, the state and county government, and the new chairman of the ACCD concerning the location and construction of a new convention center. This agreement signaled the beginning of a renewal of trust and cooperation between the chief actors in the city and partial restoration of the public-private partnership.

RENAISSANCE II

In 1976 Peter Flaherty left Pittsburgh for Washington and a post as President Jimmy Carter's assistant attorney general. His place was taken by Richard Caliguiri, former city council president and an opponent of Flaherty's in the 1973 election. Caliguiri's prime agenda item as mayor was to restore the public-private partnership and to launch Renaissance II, indicative once again of a major change in management philosophy. But the fiscal and personnel reforms of Mayor Flaherty formed a solid basis for the renewed development focus. By 1977, too, new service-sector growth was taking place and the tightening of the market for high-grade office space in the central business district was becoming quite evident.

Renaissance II shared many similarities with Renaissance I. As in 1945, many Pittsburgh leaders felt that the city was in crisis. Population, for instance, had continued to decline in spite of Renaissance I and the number of *Fortune* 500 companies headquartered in Pittsburgh had been reduced from 21 in 1960 to 14 in 1980. Yet just as in the 1945–1952 period, an extreme shortage of high-grade office space had developed, with occupancy rates reaching 99%. Faced with the combination of challenge and opportunity, the ACCD again turned its attention to planning and development on a grand scale, while the Chamber of Commerce became action oriented.

The major differences between the two renaissance periods are just as important as the similarities. Under the direction of Mayor Caliguiri, Renaissance II took on its own configuration. In Renaissance I, development goals and plans had largely been set in the private sector—by the ACCD and its planning adjuncts. In Renaissance II, the city would become a full, if not majority, partner in goal setting. A Department of Housing and a Department of Economic Development were created to set goals and operations for future growth in the city. In 1982 these departments were eliminated and the functions of housing, renewal, and economic development were consolidated into the Urban Redevelopment Authority (URA), but the city maintained its key role in development.

Like Renaissance I, an important part of the new cycle of redevelopment was high-rise buildings in the downtown. Thus, three 40-story plus office headquarters were completed by 1983, as well as a number of smaller structures. In Renaissance II, most of the market demand is met not by one massive high-rise office complex, but by a series of independent high-rise complexes, each requiring different levels of public involvement.

In Renaissance I the environment was a major issue for the public sector; it has become less important in the 1980s. The major public focus now involves major transportation improvements, for which most of the plans were actually made in the Flaherty years. Two busways into the central business district from the south and the east have begun operations while major work has proceeded on a light rail transit system from the south with a downtown subway loop. The operating date for this system was set in 1986. Additionally, through Caliguiri's initiative, the crosstown (N/S) superhighway connecting to I-79 in the northwest corridor of the region is under way.

Although the continuing improvement of the Golden Triangle has been emphasized under Caliguiri, the neighborhoods continue to receive a large share of attention and federal funds. Lower-interest loan programs, housing development, and rehabilitation and weatherization have been developed in targeted neighborhoods throughout the city. In some cases, such as on the city's north side, the partners in the renewal program included the city, neighborhood groups, and the lenders. In order to keep a finger on the neighborhood pulse, a Pittsburgh Neighborhood Alliance was sponsored by the city. Mayor Caliguiri had clearly learned the lessons of the Flaherty years.

On the private side, the dominant leadership of Richard K. Mellon had been replaced by a diversified set of CEOs, with no clearly identifiable single power center. In addition, the role of the ACCD was somewhat reduced, with the Chamber of Commerce becoming a more active organizational participant. The Chamber, for instance, played a key role in the development of loaned executive programs Committee for Progress in Allegheny County (COMPAC) to improve the efficiency of operations in the public sector. As compared with Renaissance I, in Renaissance II there was a diversity of actors, some of whom were outsiders, rather than a few giants with deep roots in the city's history.

In short, the neighborhoods have received attention and the management picture is much more diversified, but the conspicuous and spectacular high-rise office complexes in Renaissance II, like those in Renaissance I, are the key to the developmental changes. These new structures are symbolic in two respects: in the large growth of office space and employment in the central business district, and in the changing nature of Pittsburgh's economic base from one dominated by heavy manufacturing (steel in particular) to a more diversified, service-oriented base dominated by corporate headquarters. In 1957, downtown Pittsburgh had 12 million square feet of office space; in 1978 it had 19 million square feet; and by the end of 1985 it is predicted that another 6 million square feet will be added. Thus the increment in terms of office space in Renaissance II is quite large, equaling in 7 years that of the previous 20 years and symbolizing the growing importance of Pittsburgh's new service economy.

TRANSFORMATION OF PITTSBURGH'S ECONOMIC BASE

In most people's minds, Pittsburgh, steel, and environment pollution still remain synonymous. By 1880, the city was the center of the nation's largest steel-making district, although even then it was losing its comparative cost advantage to other regions. After the organization of U.S. Steel at the turn of the century, industry magnates devised a protective pricing plan (the "Pittsburgh Plus" system) to protect their Mon Valley investments—a scheme that effectively remained in place until Supreme Court decisions finally dismantled it after World War II.

Pittsburgh *was* steel and steel was dirt and pollution—hardly the environmental conditions conducive to a high-technology image! But even by the 1960s, Pittsburgh's economic base had begun to change. During the 1950s Pittsburgh was heavily dependent on durable goods manufacturing with steel accounting for nearly 20% of the region's nonfarm employment. 1957 to 1963 were years of stagnant and declining employment in all sectors except government. But between 1963 and 1979 the Pittsburgh economy recovered and began to diversify, with major growth in government, services, and trade. It was in the sharp recession of the early 1980s that Pittsburgh's economic transformation was driven home: well over 80,000 manufacturing jobs were lost in a three-year period, more than half of these in the primary metals sector, while the service sector continued its longer-term growth. The year 1980 was a significant turning point (Figure 10.1). From the specialized steel city of 1950 had emerged the services center of the 1980s, with an employment profile much like that of the rest of the nation (Table 10.1).

The employment growth was in government and in nonprofit and producer services (the complex of corporate activities that includes

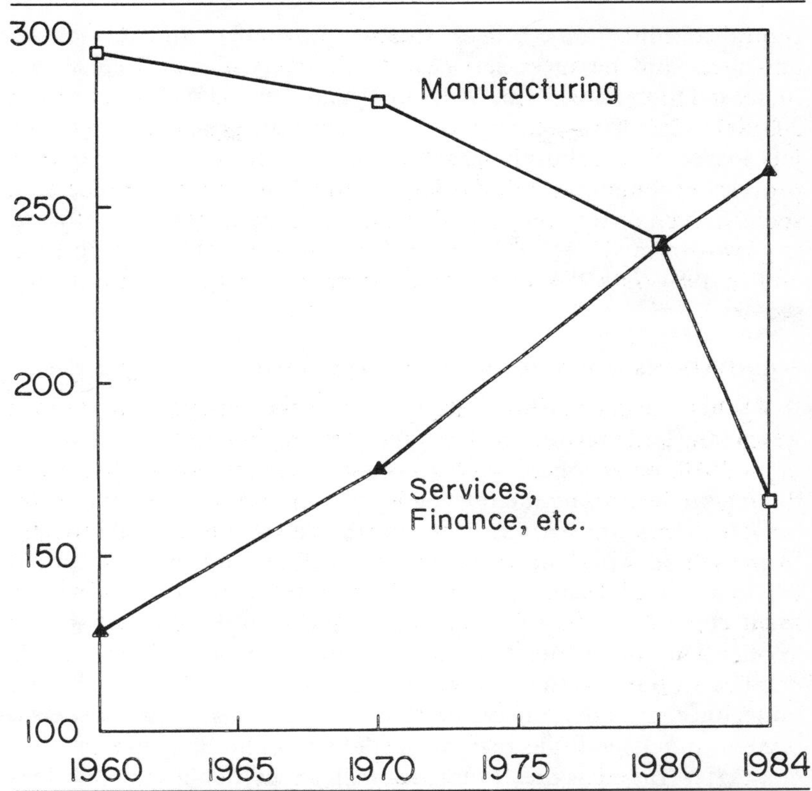

FIGURE 10.1: Employment Change: Pittsburgh SMSA (thousands)

TABLE 10.1
Percentage Goods and Service Production Employment, United States and Pittsburgh, 1952–1982

	1952	*1962*	*1972*	*1982*	*1982/1952*
U.S.					
Goods	41.4	36.8	32.1	26.7	-14.7
Services	58.6	63.2	67.9	73.3	14.7
Pittsburgh					
Goods	50.4	41.8	35.7	27.5	-22.9
Services	49.6	58.2	64.3	72.5	22.9

NOTE: Goods production is composed of manufacturing, mining, and construction. Service production consists of employment in transportation, public utilities, wholesale and retail trade, services, and government.

central administrative offices; finance, insurance, and real estate; and legal and business services). Collectively, these three sectors increased their share from 27.4% to 41.2% of the labor force, or from 203,000 to 392,000 jobs, the gains far outstripping the manufacturing job losses as Pittsburgh was transformed from a raw materials processing center to a full participant in an information processing society. Even during the recession total employment had continued to grow—from 870,902 in 1970 to 938,453 in 1980 to 1,035,500 in 1984, and in 1984 only 18% of the labor force was in the manufacturing sector.

FOUNDATIONS FOR A HIGH-TECHNOLOGY FUTURE

As this transformation of employment was occurring, foundations were being laid that outline the shape of the region's future: on the eve of the 1980s recession, there was a sharp rise in foreign investment in the region, beginning the internationalization of the regional economy; the recession years saw the growth of a venture capital industry in the region where none had existed before; and the new capital began to stimulate spin-offs from the region's research and development centers, with two emphases in particular—computer-based applications of artificial intelligence to robotics and to software engineering, and in the biological sciences and medicine, including biotechnology and biomedical engineering. And as the pace of development quickened, the region's leadership came together to chart *Strategy 21*, billed as the "Pittsburgh/Allegheny Economic Development Strategy to begin the 21st Century."

Entry of foreign capital into the Pittsburgh region accelerated after 1975, due in part to the promotional efforts of an Allegheny Conference spin-off organization, Penn's Southwest Association. In the period 1975 to 1984, 55 foreign firms chose to locate in Pittsburgh, initially creating 7700 new jobs. What is important is that many of these were the North American headquarters of their firms. Much of the interest was in new ways to conduct the traditional resource-based economy—in applications of robotics in the mining industry, for example—but a Pittsburgh base also was being established by many foreign multinationals.

Venture capital is, of course, an important source of funds for the development and growth of new enterprises, particularly high-technology firms. Originally restricted to a few families with immense personal wealth, the venture capital industry now is made up of numerous players with diverse backgrounds and organizations. The role of venture capital in both stimulating and localizing high-technology regional development is not well understood. However, it is apparent that there are marked regional differences in both the location of venture capital firms and in the dispersion of funds, and that regions experiencing the most ebullent high-technology growth are also the regions with the greatest concentration of venture capital companies. The emergence of a local venture capital industry is therefore taken by many to be a bellwether indicator of future economic growth.

In 1984 there were 10 venture capital firms in Pittsburgh with an aggregate of approximately $209 million in capital under management (Table 10.2). Most of the firms were new and small. The majority had gotten under way in the period 1981–1984.

The Hillman Company is the oldest and by far the largest venture capital firm in Pittsburgh with an estimated $100 million under management, but Hillman had traditionally not invested in Pittsburgh. The local investment is now coming from (1) venture capital firms held by corporations and (2) private venture firms. The average amount of capital under management by members of the National Venture Capital Association at the end of 1981 was $41.7 million. In contrast, in 1984 Pittsburgh's corporate and private venture firms (excluding Hillman) had an average of $17.1 million and $7.3 million under management, respectively. Their investments were, however, focused on high-technology spin-offs from the region's universities and research centers.

Because of the strong identification between high technology, economic growth, and new job creation, states are becoming concerned with attracting new high-tech firms. Pennsylvania has initiated an important new program designed to promote the development of high technology efforts in the state—the Ben Franklin Partnership for Advanced Technologies—and this, too, is beginning to have significant effects on Pittsburgh. The program is designed to foster industrial seedbeds, creating conditions conducive to the development of small entrepreneurial firms. The Ben Franklin Partnership provides state funding to Pennsylvania's research universities, other institutions of higher education, business, labor, and other groups and organizations through four regional Advanced Technology Centers. Each center provides for joint research and development efforts in concert with the private sector, in specified high-technology areas. A second major thrust of the partnership is to assist institutions of higher education in providing training and retraining in technical

TABLE 10.2
Venture Capital Firms in Pittsburgh, 1984

Firm	*Amount of Capital Under Management ($million)*
+ Hillman Company	$ 100
# Kopvenco	27
+ Fostin	20
* Security Pacific	20
+ Trivest	20
# Venwest	12
* PNC Venture Capital Group	6.5
+ C & L	1.7
+ Robinson	1.5
+* PFDC	
Total	$ 208.9

NOTE: + private venture firm; * small business investment corporation; # corporate venture firm.

and other skill areas essential to assisting first expansions and start-ups. Third, the grants are designed to assist in providing entrepreneurial assistance services, for example, linking R&D, entrepreneurs, venture capitalists, and other financial resources including assistance in preparing business plans and feasibility studies. In 1983–1984, $10 million was appropriated for the partnership and was matched by over $28 million from the private sector. Increased levels of funding are projected in the years ahead. The Western Pennsylvania Advanced Technology Center, which consists of Carnegie-Mellon University, the University of Pittsburgh, and 11 other public and private colleges and universities and several private firms, has focused its research efforts on robotics and computer-aided manufacturing, and biotechnology.

In 1984 Pennsylvania passed additional legislation making more money available for encouraging new businesses and enhancing the pool of venture capital. The primary element of that legislation was a $190 million bond issue creating the Small Business Incubators Act. This act is administered by the state board of the Ben Franklin Partnership Fund; $20 million has been earmarked for grants to the Advanced Technology Centers, $3 million of which is to assist in establishing seed money funds. These funds will be used to match money from the private sector with the state providing an additional $1 for every $3 of private sector investment in new enterprises. The remaining $17 million is earmarked for loans for establishing, operating, and administering small business incubators. Incubators are not permitted to include retail, not-for-profit, or wholesale enterprises or personal service corporations.

In addition to the Small Business Incubators Act, other legislation guides the spending of the remaining $170 million—such as $50 million for business infrastructure improvements, $27 million for vo-tech equipment in schools, and $15 million to assist in employee ownership of troubled businesses. Moreover, the State Employees Retirement System and Public School Employees Retirement System have been allowed to invest up to 1% of their total pension fund assets in venture capital firms. The assets of the two pension funds total nearly $9 billion.

Penn's Southwest Association reports that the Pittsburgh region now has some 40 major R&D centers with some 25,000 professional employees. A major increment will be the Defense Department's newly established Software Engineering Institute, to be managed by Carnegie-Mellon University. In five years, SEI will create 250 high-level jobs and operate with a $33 million annual budget. But more important, an industrial affiliates consortium of more than 100 firms has been created to provide supplementary funding, and many of them have indicated that they will be establishing branches close to SEI: a "thoughtware" agglomeration typical of modern producer-service economies is in the making. Similarly, the Gulf Oil Company's research facility at Harmarville, closed after Gulf was taken over by Standard Oil of California, was given to the University of Pittsburgh and will be used to house the first round of biotechnology spin-offs from that institution. The expectation is that it will have the same kind of catalytic effect as SEI will have within Pittsburgh's Oakland community.

Strategy 21 is a plan by the City of Pittsburgh, Allegheny County, and Carnegie-Mellon University and the University of Pittsburgh to engage the Commonwealth of Pennsylvania in further efforts to promote the changes under way within the region. The view of the future is best expressed by quoting directly from the planning document (p. 1):

> While the region has been known in the 20th century as a steel-making giant and a corporate headquarters capitol, the . . . economy of 21st century Pittsburgh must be positioned to take maximum advantage of

emerging economic trends toward advanced technology and international marketing and communications systems. A diversified economic base must be created that includes light as well as heavy manufacturing, that capitalizes on the region's natural resources, and that promotes a new mix of large and small businesses marked by a renewed spirit of entrepreneurship and university-linked research and development. Such an economic base will provide the region with optimum flexibility to move in new directions as growth opportunities appear and will give its people much greater choice in how they spend their working lives... The Pittsburgh/Allegheny partnership proposes a strategy to reach the following goals:

1. reinforce the region's traditional economic base as a center for the metals industry and an international corporate headquarters;
2. convert underutilized land, facilities and labor force components to new uses, especially those involving advanced technology;
3. enhance the region's quality of life, thereby attracting new residents and increasing tourism; and
4. expand opportunities for women, minorities, and the structurally unemployed.

In pursuance of these goals, the Mayor, the County Commissioners and the University Presidents propose to expand their regional partnership to include the Commonwealth of Pennsylvania by advocating that future state-funding focus on five specific project areas.

These areas are as follows: airport development to improve the region's international access; transformation of underutilized land in the ring surrounding the areas already redeveloped in Renaissance I and II, including significant riverfront improvement; redevelopment of the Mon Valley mill towns, where the problems of dislocation caused by the collapse of the steel industry are most intense; transportation improvements within the region; and enhanced support for the Western Pennsylvania Advanced Technology Center. To quote *Strategy 21* again (p. 2):

The advanced technology research projects being developed by the University of Pittsburgh and by Carnegie-Mellon University would have the multiple effect of (1) assisting in the revitalization of the metals industry through the introduction of new technology; (2) strengthening the region's attraction as a corporate headquarters by increasing the local availability of research and development expertise; and (3) capitalizing on the comparative advantage that research efforts of the Pittsburgh university community have over those of other competing universities, as demonstrated by the awarding of the $103 million Software engineering Institute contract to Carnegie-Mellon University. The synergistic effect of these research projects would advance the economic revitalization of the entire region.

It is proposed that these projects be funded by both public and private funds. The initial proposal to the Commonwealth of Pennsylvania calls for raising $425 million in state money. Private sector investment of nearly $1.1 billion expected to pay more than half of the cost of carrying out the diverse projects with state, federal, and local government funds to pick up the remainder.

SOCIAL AND GEOGRAPHICAL CONSEQUENCES OF THE CHANGING ECONOMY

Even before the final collapse of steel after 1980, changes in the region's economy were having profound social and economic consequences, the other side of the transformation. These may be sketched in a series of tables.

Between 1970 and 1980 manufacturing was declining and producer services and the nonprofits were growing (Table 10.3). The earlier manufacturing economy had produced a blue-collar region, for employment was heavily concentrated among the craft workers and operatives, in contrast to the professional/clerical employment structure of the newer-growth industries (Table 10.4).

Thus, as the structure of the regional economy shifted, so did its occupations (Table 10.5). Blue-collar jobs shrank and both professional and managerial and sales and clerical occupations grew.

One profound effect has been in the income distribution: manufacturing jobs created a blue-collar middle class; but earnings in the new-growth industries are biased to the top or the bottom of the wage distribution (Table 10.6). The manufacturing-based middle class has been shrinking, and job growth has been in the top or the bottom earnings categories (Table 10.7). A region whose lifestyles were dominated by a blue-collar middle class is being transformed into one in which there is increasing inequality between a well-paid professional and managerial elite and large numbers of close to minimum-wage clerical and service workers.

The effects have been asymmetric by race and sex. First, there are significant differences in who are employed in which occupations

TABLE 10.3
Employment Change by Industry: Pittsburgh SMSA, 1970-1980

	(Numbers Employed)				
	1970	*(%)*	*1980*	*(%)*	*1970-1980 Change*
Extraction	13,729	1.6	16,943	1.8	+ 3,214
Construction	45,811	5.3	49,691	5.3	+ 3,880
Manufacturing	275,989	31.7	239,937	25.6	- 36,052
Distribution	96,358	11.1	112,135	11.9	15,777
Retail	143,059	16.4	160,548	17.1	+ 17,489
Consumer services	46,669	5.4	44,940	4.8	- 1,729
Producer services	92,811	10.7	124,860	13.3	+ 32,049
Nonprofits	118,159	13.6	157,703	16.8	+ 39,544
Government	35,317	4.0	31,696	3.4	- 3,621
Unreported	3,000	0.3	N.A.	—	—
Total	870,902		938,453		+ 67,551

TABLE 10.4
Occupational Distribution of the Labor Force in Selected Industries in Percentage: Pittsburgh SMSA, 1970

	Manufacturing Industry	*Retail Services*	*Producer Services*	*Non-profits*	*Govern-ment*
Professional/technical	11.7	2.3	27.0	51.6	10.4
Managerial	4.6	12.4	9.3	3.4	9.2
Craft workers	23.6	7.3	3.1	2.1	4.2
Operatives	33.3	10.1	2.1	1.6	2.7
Sales Employees	2.7	29.2	10.7	0.2	0.4
Clerical workers	13.8	15.4	34.7	15.0	42.0
Service workers	2.6	18.7	12.1	25.6	26.9
Laborers	7.6	4.6	1.2	0.5	4.2

(Table 10.8). This has meant that the employment shifts by occupational categories have produced differential benefits by race and sex: white males have been the principal beneficiaries of professional and managerial job growth; females generally have found their employment in typically sex-labeled low-paying occupations (Table 10.9)—indeed, there is a double irony. Black males had been finding their

TABLE 10.5
Employment Change by Occupation: Pittsburgh SMSA, 1970-1980

	Numbers employed				
	1970	*(%)*	*1980*	*(%)*	*1970-1980 Change*
Professional/technical	137,318	(15.8)	156,996	(16.5)	19,678
Managerial	63,328	(7.3)	92,876	(9.8)	29,548
Craft workers	134,848	(15.5)	126,644	(13.3)	- 8,204
Operatives	145,519	(16.7)	123,389	(13.0)	- 22,130
Sales employees	69,202	(8.0)	94,519	(9.9)	25,317
Clerical workers	157,918	(18.1)	166,144	(17.5)	8,226
Services workers	111,037	(12.8)	132,066	(13.9)	21,029
Laborers	47,371	(5.4)	52,986	(5.6)	5,615

TABLE 10.6
Distribution of Employment by Earnings Category and Industry in Percentage: Pittsburgh SMSA, 1970

	Earnings Category		
	High	*Medium**	*Low*
Extraction	8.6	43.3	48.1
Construction	13.6	63.1	23.3
Manufacturing	16.3	59.6	24.0
Distribution	16.7	48.7	34.6
Retail	14.7	46.6	38.7
Consumer services	10.9	28.3	60.8
Producer services	36.3	15.8	47.8
Nonprofits	55.0	3.8	41.2
Government	19.6	7.3	73.0

* ±15% of the national average wage.

greatest employment opportunities in the well-paying craft and operative jobs in manufacturing industry at the very time that this economic base collapsed. The alternatives are much lower-paying service jobs (compare Tables 10.8 and 10.10).

Thus, black poverty levels have remained extremely high in the region (Table 10.11). Collapse of the manufacturing base has surely been one of the reasons for the rapid increase of full-time female

TABLE 10.7
Employment Change by Earnings Category And Industry: Pittsburgh SMSA, 1970–1980

	Earnings Category		
	High	*Medium*	*Low*
Extraction	276	1,392	1,546
Construction	528	2,448	904
Manufacturing	-5,882	-21,509	-8,661
Distribution	2,635	7,683	5,459
Retail	2,571	8,150	6,768
Consumer services	-187	-492	-1,050
Producer services	11,646	5,069	15,334
Nonprofits	21,749	1,502	16,292
Government	-711	-265	-2,644
Net change	32,625	3,978	33,948

TABLE 10.8
Occupational Distribution by Race and Sex, in Percentage: Pittsburgh SMSA, 1971–1975

	White Males	*Black Males*	*White Females*	*Black Females*
Professional/technical	15.8	7.1	16.2	12.7
Managerial	13.6	4.2	4.6	2.5
Craft workers	21.7	12.7	1.0	0.8
Operatives	26.0	35.6	6.8	8.7
Sales employees	4.7	2.8	13.5	9.5
Clerical workers	5.3	4.3	40.5	33.0
Service workers	4.0	16.2	13.6	27.5
Laborers	9.0	17.0	3.6	5.3

labor force participation in the region (Table 10.12). Loss of manufacturing jobs has forced women back into the minimum-wage services work force in the attempt to maintain household incomes.

Loss of manufacturing employment in the region has had other negative impacts on older blue-collar communities. Allegheny County lost 9.7% of its population between 1970 and 1980 as young workers

TABLE 10.9
Employment Change by Earnings Category, Race, and Sex: Pittsburgh SMSA, 1970–1980

	Earnings Category			
	High	*Medium*	*Low*	*Totals*
White males	27,784	3,237	13,739	41,760
Black males	608	190	1,701	2,499
White females	6,769	508	16,726	24,003
Black females	463	43	1,782	2,288
Totals	32,625	3,978	33,948	—

TABLE 10.10
Earnings Differentials by Race, Sex, and Industry: Pittsburgh SMSA, 1975

	White Males	*Black Males*	*White Females*	*Black Females*
Extraction	1.11	D	0.47	D
Construction	0.96	0.63	0.41	D
Manufacturing	1.15	0.84	0.58	0.54
Transportation, communications, utilities	0.96	0.80	0.63	0.63
Wholesale/retail	0.75	0.44	0.30	0.24
Finance, insurance, real estate	1.02	0.58	0.48	0.44
Other services	0.86	0.56	0.42	0.43
Government	0.73	0.62	0.44	0.44

SOURCE: 10% continuous work history sample.
NOTE: Average earnings of all white males in all industries is set equal to 1.0; D=deleted.

migrated South, dropping from 1.6 to 1.4 million people. This decline has been more significant in some areas than in others. For example, 22.4% of the population in the less affluent blue-collar communities have left.

On the other hand, well-paying professional job growth is having the opposite effect on other inner-city neighborhoods. Gentrification

TABLE 10.11
Income Below Poverty Level: Pittsburgh SMSA, 1980

	Total		*White*		*Black*	
	1970	*1980*	*1970*	*1980*	*1970*	*1980*
Total families	621,432	615,915	582,483	—	38,949	—
Families below poverty	44,782 (7.2)	39,516 (6.0)	34,873 (6.0)	28,532 (5.0)	9,909 (25.4)	10,529 (24.5)
Female household heads	18,980	21,044	12,717 (37)	12,710 (45)	6,263 (63)	8,147 (77)
With children under 18	15,277	18,927	9,551 (27.4)	11,087 (39)	5,726 (58)	7,674 (73)
65 years and older		108,199 (18)				
Unrelated individuals below poverty	68,056 (40)	60,726 (24)	58,843	49,436 (22)	9,213 (50)	10,702 (41.5)
65 and older	(52.2)	(37.1)		19,055	(42.1)	3,393 (32)
Persons below poverty	225,706 (9.5)	193,047 (8.7)	175,431 (7.8)	142,773 (7.0)	50,275 (30.2)	48,190 (28.2)
65 and older	(24.9)	30,025 (15.6)	—	25,378 (18)	(12.7)	4,537 (9.4)

TABLE 10.12
Selected Characteristics of the Pittsburgh Labor Force, Percentage 1960–1980

	1960	*1970*	*1980*
Percentage of labor force female	29.3	34.0	42.7
Female labor force participation rate	28.6	34.3	40.0
Percentage of female work force			
Working part-time	24.9	30.5	22.7
Working full-time	66.1	61.5	74.3

in the Pittsburgh region had been limited compared to other major cities in the country, but after 20 years of relative inactivity the process is accelerating. Playing a pioneering role, Pittsburgh History and Landmarks Foundation had been responsible for limited-scale preservationist-gentrification of two small, early nineteenth-century neighborhoods (the Mexican War Streets and East Carson Street on the South Side), but the scale of change was very small until the full force of the "yuppies" came to be focused by the private market in Shadyside and adjacent areas, close to Oakland's university, medical center, and high-tech complexes, where rehabilitation and condominium conversions proceed apace.

The result in the city is a patchwork of rising and falling neighborhoods, responding to the economic shifts that are polarizing the social structure. This transformation of the social life and of the geography is a reality to which the region's partnerships have yet to respond. Increasing militancy in the Mon Valley mill town communities is evidence enough that Rand McNally's ratings failed to ask "livable for whom?" High-tech Pittsburgh is ascendant, but for blue-collar Pittsburgh opportunity seems progressively more remote; Pittsburgh's future history seems destined to be Dickensian—quite different, depending in which of the two cities Pittsburghers live.

11

Knowledge and the Advanced Industrial Metropolis

RICHARD V. KNIGHT

□THIS CHAPTER IS concerned with how we think about cities, particularly industrial cities, and how we envision their future. Cities, which date back over 6000 years, have, at least up until the advent of the industrial revolution, all been of a similar type, performed the same basic functions for society, been built in a similar way, and have taken somewhat similar forms. But the industrial city, which is only now beginning to be recognized as a new type of world city, is founded on a different set of values and its development is taking forms that are very different from those of traditional cities. Although the same general philosophical framework that is used for thinking about traditional cities can be used for thinking about industrial cities, certain distinctions have to be made between these two types of cities if we are to increase our understanding of what is, in the context of the history of cities, a new type of world city.

Industrial cities differ from traditional cities in some very significant ways that will be explored, but these differences are difficult to discern because the values that underlie industrial cities have not yet been fully articulated in the form of a city. Industrial cities are still in the initial stages of being established; only their institutional base has been formed—they have yet to be built in a physical or cultural sense. It is important that we increase our understanding of the factors underlying these differences so that industrial metropolises will be able to capitalize on their particular advantages as they build industrial cities. If these differences are accounted for, the

---

AUTHOR'S NOTE: *This chapter presents material from a much larger work currently under preparation and represents part of a symposium conducted in Berlin in late 1984.*

building of the industrial cities can provide a framework for accommodating the structural changes that are occurring in the economies of industrial metropolises.

The building of an advanced industrial city provides industrial metropolises with a means for sustaining their development while communities in the region undergo the transition from manufacturing to knowledge centers. The building of the industrial city, by accommodating the expanding knowledge-intensive functions, will serve to offset dislocations of manufacturing activities. Industrial metropolises, though losing their comparative advantages vis-à-vis manufacturing activities, have also been establishing new advantages related to knowledge-intensive activities. But the advantages that have accrued as a result of the expansion of knowledge-intensive activities are of temporal nature and could easily be lost if communities are unaware or unsupportive of them. To retain their knowledge resources, industrial metropolises will have to upgrade their environments and build world-class cities that can compete effectively with traditional cities for knowledge resources in the global economy.

In short, with the advent of the global economy, the role of the advanced industrial metropolis is changing; manufacturing centers are becoming knowledge centers. As technology continues to be advanced and as instantaneous worldwide communications become more widespread, industries will be further rationalized and restructured on an international basis. This application of advanced technology and the concomitant restructuring of industries is driven by powerful global forces that are very difficult to shape. As a result, the comparative advantage of industrial metropolises is changing. Their locational advantages have been eroded and manufacturing activities are declining. At the same time, however, knowledge-intensive activities have continued to expand, and as a result, regions that initially depended on manufacturing activities for an economic base are becoming more and more dependent on knowledge-intensive activities. It is on the continued development of these knowledge resources that the evolution of industrial cities depends.

NEW INDUSTRIAL DEVELOPMENT POLICIES

It is important to note that these newly founded knowledge resources are human-made and that they are likely to be lost unless development strategies designed to strengthen them are implemented. This would require a major reevaluation of industrial development policies since most industrial development efforts are geared toward retention of manufacturing activities. They do not succeed because locational advantages, once lost, cannot be regained. Natural advantages such as locational advantages can be exploited by transportation and infrastructure improvements, but when they shift—and shifts are brought about by a complex set of forces such as technological advances, the development of new sources of energy, raw materials, of new markets, and so on—there is little that can be done to regain them. Manufacturing activities have always gravitated toward locations where production costs can be minimized. Whenever locational advantages shift, companies have to relocate manufacturing operations in order to remain competitive. Industrial development policies have a long tradition of promoting production activities and exploiting locational advantages. Consequently, there is a great reluctance to concede that manufacturing activities will continue to decline or that locational advantages may dictate their relocation away from industrial metropolises.

Industrial development strategies geared to knowledge-intensive activities offer considerable potential because knowledge resources are human-made and depend on advantages that can be created at the metropolitan or regional level. Knowledge resources take many different forms; they are created by a wide variety of advanced industrial, scientific, commercial, and cultural organizations. Although the presence of knowledge-intensive activities in industrial metropolises is in most cases a historical accident, they can be retained if appropriate development strategies are implemented. Knowledge resources represent a distinct advantage in advanced industrial society because as technology advances, the knowledge content of goods and services produced increases. More attention needs to be given to the contribution of knowledge resources in wealth creation. Industrial development policies in advanced industrial metropolises should thus become more concerned with how and where knowledge is produced. The value added by workers engaged in knowledge-intensive activities is increasing, whereas value added by production workers in manufacturing activities is declining.

Knowledge-intensive activities such as corporate administrative and auxiliary activities and related technical, scientific, professional, managerial, specialized business, and cultural activities may have begun as appendages of manufacturing operations, but in recent decades they have become resources in their own right. As industrial firms have expanded their operations worldwide, activities that began as support services for locally based manufacturing plants have been upgraded and expanded in order to serve these growing worldwide operations. In contrast to manufacturing operations, which have been declining in industrial metropolises over the last two decades as manufacturing has been relocated elsewhere, knowledge-intensive activities have grown steadily and, as a result, have become increasingly dominant. This legacy of knowledge resources from the industrial and scientific revolutions now provides industrial metropolises with a knowledge resource base on which they could establish a very special niche in the global economy.

These new advantages should not be taken for granted because as the role of knowledge in industrial production increases, competition for knowledge resources will become more and more intense. Although many advanced industrial metropolises presently have large concentrations of knowledge-intensive industrial activities, they are there due to a historical accident. Industrial metropolises can capitalize on their knowledge resources by creating environments that are conducive to the expansion of knowledge-intensive activities. In order to maintain their position in global markets, advanced industrial organizations have to be able to recruit talent from around the world. To do this they have to offer a competitive quality of life. Strategies that would enable industrial metropolises to secure their knowledge base and sustain their development as knowledge centers should be carefully considered as part of a comprehensive development strategy. Conventional industrial development policies aimed at shoring up declining manufacturing activities should not be abandoned but should be reevaluated in the context of a broader and longer-term development strategy, that is, a strategy that incorporates knowledge-based activities.

A NEW TYPE OF WORLD CITY

The prospect of building a new type of world city is not simply idle speculation; it is conceived of as a strategy for sustaining the development of industrial metropolises and offsetting the declines that are occurring in manufacturing activities throughout industrialized nations. It is a strategy for conserving the human and cultural resources that have been carefully nurtured there over the last century or so but that could be lost within a decade unless action is taken to secure them. Knowledge-intensive activities have become an integral part of the advanced industrial economy and now need to be woven into the fabric of the industrial metropolis. To do this, we will have to increase our understanding of the role that knowledge plays in advanced industrial and scientific activities and build communities that are responsive to the needs of organizations and persons engaged in these activities.

The advanced industrial city must be allowed to take a form that honors the cultural, psychological, and physical needs of advanced industrial and cultural activities. The industrial city is unlikely to fit the mold of traditional cities; it will be shaped by different forces and by values that will surface as the industrial and scientific revolutions progress. These values may be tempered by traditional values but they will, in all likelihood, be of a different nature. Each city should be allowed to take whatever form is appropriate to its particular situation.

Unfortunately, it is unlikely that industrial metropolises will consider the option of building world-class cities based on their knowledge resources unless there is a national city policy that encourages such efforts. If, alternatively, the building of these prototype cities is left to occur in an ad hoc fashion, many regions, particularly those that are in decline, will forfeit the opportunity by default. The social costs of not building these cities could be tremendous. Building the industrial city is a challenge that no advanced industrial nation can afford to overlook. Those regions and nations that seize this unusual opportunity will in the future be looked back upon as contributing to a major milestone in the history of cities.

The process of building the industrial city is, in fact, well under way. The first step, the institutionalization of new values created by the industrial and scientific revolutions has already occurred at the organizational level. Many industrial and nonindustrial corporations, research and medical-university centers have already established firm footholds in very competitive global markets. But the next step involves institutionalization of these values at the community level. It is ironic, but the process of gaining a greater appreciation for and a broader acceptance of the values that underlie industrial and scientific society will be more difficult in the industrial metropolises than in areas where these organizations are not presently based.

Industrial regions have been jaded; they developed accidentally on the basis of locational advantages and have inculcated attitudes and behaviors that are not conducive to the building of industrial cities, which must be intentional. If industrial cities are to compete successfully with traditional cities for knowledge resources, they will have to do so on the basis of their human-made advantages, namely on their human and cultural resources and the quality of life that they offer. If industrial regions do not build intentional cities, they will lose their advanced industrial activities, and will not be able to offset the declines that are occurring in manufacturing activities throughout industrialized societies.

Existing world-class cities are referred to here as traditional cities because they have been built on the moral order, as opposed to industrial cities, which are being built on the newly emergent scientific order. What makes industrial cities so different is that they are not based solely on values and traditions that have been forged over centuries of political and religious strife but are now widely accepted and unchallenged as the moral order. They must also contend with and help to forge the relatively new and little understood values and traditions of the industrial and scientific order.

In spite of the values that the industrial and scientific revolutions have contributed to humanity, the institutions that are founded on them are generally mistrusted, even by communities that in the long run stand to benefit most. These institutions are mistrusted because of the nature of changes that have occurred over the last few decades, namely plant dislocations and the expansion of headquarters-related activities, which are not well understood. This distrust is unfortunate because the presence of advanced industrial and scientific organizations and related advanced service and cultural activities is what provides the industrial metropolis with an institutional base for building the industrial city.

Centers of advanced industrialized activities are not generally appreciated because their present and potential contribution to development is overshadowed by the problems created by plant dislocations. Their growing role as knowledge centers is not and will not be perceived as long as industry continues to be defined in terms of manufacturing activities alone. Advanced industrial and scientific activities have been overlooked because they usually begin in a small and inconspicuous manner and they evolve gradually as science is advanced and new knowledge is gained through a long process of applying technological and scientific advances to manufacturing processes.

The fact that most industrial metropolises are presently losing factory jobs and population tends to obscure the steady upgrading and expansion of advanced industrial and scientific activities. This is most unfortunate, because the prospects for building industrial cities and retaining these knowledge-intensive activities in industrial metropolises are contingent upon an appreciation of the important role that they play in the global economy.

Clearly, what characterizes an industrial metropolis as an imminent world class city is not the presence of manufacturing activities per se, as symbolized by smokestacks and factories around which these regions originally grew, but the presence of science-based organizations that contribute the knowledge inputs for manufacturing facilities now being built around the globe. As science advances and the contribution of knowledge to advanced industrial production increases, the actual value added on the production line by unskilled workers decreases. Concomitantly, the value added by knowledge workers, whether on-line by highly skilled production workers or off-line by technical, managerial, or professional workers, increases. In short, as the global economy expands and manufacturing operations are expanded around the world, the demand for knowledge-intensive activities needed to support these manufacturing activities increases.

Advanced industrial metropolises should be viewed in terms that account for all the value that is being added there today. The contribution of knowledge resources, human and cultural resources, and the broad array of specialized production and knowledge-support services should be accounted for explicitly. If industrial metropolises continued to be perceived in terms of how value used to be added, that is, in terms of the number of unskilled workers engaged in factory production jobs or the tons of goods and materials produced or shipped locally, they will be seen as declining rather than as being in transition. In short, value created in industrial metropolises continues to increase but its form is changing. The concept of the industrial city is used here as a euphemism for the potential transition of manufacturing centers into knowledge centers and for the cultural transformation of industrial metropolises into world cities.

THE ADVANCED INDUSTRIAL CITY

What makes it difficult to define the industrial city is the fact that it is based on a new form of power, namely industrial and scientific know-how. Institutionalization of this power, although well advanced at the organizational level (notably, international corporations, both industrial and nonindustrial, that is, research centers, and university-medical complexes) is only just beginning to occur at the community level. Industrial metropolises are often referred to as industrial cities even though the organizations that make up their institutional base have not organized or used their power to build such a city. A world-class city will not happen if we wait for it; the building of a world-class city will have to be a willful act. And before this can occur there must be a vision of what such a world city will be like.

The concept of an industrial city has to be translated into an achievable goal. Industrial cities will not develop accidentally as manufacturing centers did (by exploiting natural or locational advantages); major commitments will be required in order to upgrade institutions and attributes of the environments on which knowledge-intensive activities depend. Advanced industrial and scientific activities, unlike the factories from which they evolved, are dependent upon their environment for support. If their needs are not accommodated, they will relocate. Whenever this happens, the institutional or power base upon which the industrial city could be built erodes. If the city-building process is not begun immediately, industrial metropolises may find that they have lost the opportunity.

Industrial know-how is a new form of power that has yet to be articulated in the form of a city; this is why it is so difficult to perceive it and why its presence is so tenuous. Just because an industrial metropolis happens to have a strong institutional base does not mean that these organizations will remain or that they can be securely anchored. These advanced industrial activities may be present in a functional sense but they are usually absent in a cultural sense. Although these activities require and in most cases could sustain a cosmopolitan environment, the cultural and institutional orientation of these older industrial regions tends to remain aligned to the needs of the manufacturing activities and unskilled workers around which they originally formed. The built environment and the general ambience of the area continues to reflect values and needs of resource-based manufacturing activities. The industrial city per se has yet to be built.

It is interesting to speculate on how industrial cities will look when they are built. Indications are that they will be very different from traditional cities, both in respect to the cultural values that underlie them and in the way that these values are articulated in aesthetic, psychological, and spatial terms. Industrial cities will be shaped by the same forces that drive the industrial and scientific revolutions.

With the advent of the communications and information age, we may see cities take new forms.

Although many industrial metropolises have the potential for developing into world-class cities, the important question is whether these opportunities will be perceived or sought. A strong institutional base is essential but not sufficient to overcome the problems inherent in the transition of a manufacturing center into a knowledge center. A great deal of vision and leadership will be required if industrial metropolises are to capitalize on their industrial heritage and build world-class cities. They have to accomplish in a few decades what it has taken traditional cities centuries to establish—an identity, a sense of place, a civic culture, and a sense of their particular form of power and destiny.

Industrial metropolises could easily lose those organizations that could serve as an institutional base for a world city. This is because knowledge-intensive activities of advanced industrial, scientific, commercial, and cultural organizations must continuously advance their capacities and thus are consistantly outgrowing the communities where they are based. Knowledge and science-based industrial activities require very different kinds of environments and services from those required by labor-intensive manufacturing activities from which they evolved. For example, the working-class neighborhoods that grew up accidentally around mills and factories early in the industrial revolution are not well suited for workers in today's advanced industrial activities. Older industrial areas that have become blighted and dysfunctional need to be redesigned and rebuilt in a way that meets today's standards. Similarly, schools and other services also need to be upgraded. Individuals and organizations engaged in advanced industrial activities are under constant pressure to improve their performance; they cannot survive in communities that are not responsive to their needs.

In order to build an industrial city the metropolis must be governed effectively. It must have autonomy, that is, control and command over its domain, whether it is of a territorial or scientific nature. Its stature is determined by the extent of its influence. To build a world city, a metropolis must have sufficient power and autonomy to command and control resources that are critical to internationally oriented organizations that are based there. Its success at governance determines its future because this is how it maintains its power base. Cities are, by definition, elitist institutions, and to understand the differences among cities, we must take into account differences in the composition of and in the values that underlie elites in different types of cities.

The problem of understanding industrial cities is related to what C. P. Snow calls the problem of two cultures—the splitting up of the intellectual and practical life in Western society into two polar groups, the traditional and the scientific cultures. The industrial and scientific culture is not well understood or appreciated by the literary intellectuals and traditional culture is not understood or appreciated by the scientists. "Between the two (poles) is a gulf of mutual incomprehension—sometimes hostility and dislike, but most of all, lack of understanding." Snow agrees that "the scientific culture is a culture, not only in an intellectual but also in an anthropological sense" (Snow, 1961). These differences in the cultures and in their underlying value base have a very significant effect on how we perceive and think about cities. Viewed from the perspective of traditional resource-based activities, industrial metropolises are thought of as being declining manufacturing centers, but viewed in the context of a knowledge-based society, industrial metropolises are seen as being transformed into knowledge centers with the potential of evolving into a new type of world city.

The industrial and scientific revolutions, which have given rise to this new form of power and a new type of institutional base for cities, are now sufficiently advanced that the "capitals of technology" can be established. This new type of world city is, however, still difficult to discern because of the worldwide restructuring of industrial activities that is now under way. This restructuring is bringing about some profound changes in the economies of advanced industrial countries, particularly in areas where manufacturing was initially concentrated. As manufacturing is rationalized at the international level, industrial organizations are moving the less skilled and more labor-intensive operations away from industrialized and developed areas to less developed areas. Concomitantly, they are upgrading and expanding the more knowledge-intensive functions such as administrative, research and associated professional, technical, and specialized business services that are required by the growing worldwide networks of manufacturing operations. As a result, many industrial metropolises are undergoing a transition from manufacturing to knowledge centers.

This transition is extremely challenging. The opportunities are great but there are major problems that have to be addressed if these opportunities are to be realized. However, unless the nature of the transition is well understood by the community at large, their resources will be dissipated. They will be used to perpetuate outmoded structures and programs instead of shifting them to support new structures and programs aimed at adapting to change. The choice of resisting or adapting to change is a very difficult choice for a community to make, but if it is not made, communities run the risk of forfeiting their option. It is extremely difficult for a community that began accidentally and boomed on the basis of exploiting its natural advantages to change its policies and to plan for a future based on its human-made advantages, on its human and cultural resources, and on its quality of life. If an industrial metropolis wants to maintain its knowledge resources, it has to be able to offer a quality of life that is competitive with that offered by other world-class cities.

In summary, in order to advance our thinking about city development, we have to be able to conceptualize the changes that are occurring in advanced industrial cities. This is necessary because as the global economy expands and as international trade increases, the role of the older industrial cities in the global economy will continue to change. Their role in advanced industrial society is changing; their role as centers for the production of goods is waning but their role as knowledge centers is waxing. Industrial metropolises should be expanding their role as centers for the advancement and management of industrial, commercial, and cultural know-how. Unless industrial cities increase their understanding of the nature of the transition that is occurring, they will be unable to realize their potential role as knowledge centers and build world cities that will maintain their future development.

Policymakers must begin thinking of cities in the context of developments that are occurring in the global economy. Individual cities can begin by identifying those activities that are internationally oriented or have the potential for becoming so. While examining the nature of their linkages with the international community, attributes of the environment that are critical to their continued development can be identified and strategies designed to nurture knowledge-intensive activities can be developed. It is unlikely, however, that this will be done unless they increase their understanding of the nature of wealth creation in the knowledged-based society. In short, in order to advance their thinking about the future of industrial cities, they will first have to reevaluate their worldview of how wealth is created.

Just as Adam Smith provided a conceptual framework for the expansion of the manufacturing sector and the shift of work from the field to the factory, we must now provide a conceptual framework that will account for the expansion of the knowledge sector and the concomitant shift of work from the factory to the office, that is, to administrative centers, technical centers, laboratories, hospitals, libraries, classrooms, and to a certain degree back into the home. We will not be able to have confidence in advanced industrial society unless knowledge is viewed as a factor of production, unless the knowledge worker is viewed as adding value, and unless payment is made for knowledge as it is now made for the traditional factors of production, land, labor, and capital (Knight, 1986).

Knowledge-intensive activities are not easily defined in terms of standard industrial classifications or occupational categories and should definitely not be equated with white-collar or service activities that may or may not be knowledge-intensive in nature. Nontheoretical knowledge is beginning to receive wider recognition even in traditional manufacturing production activities. In fact, a new payment system called "payment for knowledge" is being discussed in the United States by the United Auto Workers Union in their current negotiations for the proposed Saturn plant. The purpose of the scheme is to change the inflexible work rules that have inflated production wage costs and undermined the competitiveness of the industry. Under the new plan, the worker would be paid according to the number of tasks he or she can perform, thus increasing flexibility and reducing the number of workers required on line. General Motors's new Saturn plant, which will produce 500,000 cars a year, will under the proposed payment plan require 3,000 workers per shift on a two-shift workday compared to 5,000 workers required to get a similar output in 1980. By assigning workers multiple tasks, the completion and quality of which would be aided by high-technology tools, the new Saturn plant will reduce the cost of production wages.

CONCLUSION

Concerns about how an advanced industrial city actually works in an economic, social, and political sense and doubts as to its legitimacy—whether such a city can earn an "honest" living by producing and exporting knowledge—are sure to linger. Clearly, the real challenge before us is to shape these new ideas so that the new forms of wealth and development can be identified, conceptualized, documented, and made real. Although the advanced industrial city and its associated knowledge-based society is, in a functional sense, present, we lack the philosophical framework for understanding it. It cannot be fully acknowledged or appreciated until it is present in a cultural sense.

In order to increase our understanding of the advanced industrial city, the manifestations of the knowledge-based society will have to be defined at the metropolitan or city-region level. Once the particular attributes of a specific advanced industrial city have been identified, then those attributes can be enhanced in ways that are conducive to continued development. In this way the advanced industrial city will become culturally present and will be thought of in terms of its human-made resources, its human and cultural resources, its institutional base, its quality of life, and so on. In order to build a civic culture and a sense of place, each industrial metropolis must learn how to learn from its past so that the city can capitalize on its scientific contributions, on its cultural traditions, and on the values that underlie them.

An industrial metropolis, by achieving world-class status through building an advanced industrial city, will not only secure its future by improving the welfare of its citizens, but in the process it will also be contributing to economic, social, and cultural progress. The advanced industrial city will, by its nature and the conditions placed on its development, become an increasingly humanistic place.

REFERENCES

KNIGHT, R.V. (1986) "The advanced metropolis: a new type of world city," pp. 391–436 in H.-J. Ewers et al. (eds.) The Future of the Metropolis: Berlin-London-Paris-New York. New York: de Gruyter.

SNOW, C.P. (1961) The Two Cultures and the Scientific Revolution. New York: Cambridge University Press.

12

The Arts and Urban Development

WILLIAM S. HENDON and DOUGLAS V. SHAW

Eenie, Meenie, Minie Mo
Catch a Yuppie with a show
Even where there's ice and snow
Towns with culture ought to grow.

□TRADITIONALLY CITIES HAVE grown through a combination of high fertility, immigration, and the steady depopulation of the countryside. In recent decades, with lower birth rates, immigration sharply curtailed, and a new demographic balance struck between city and farm, urban growth as a mere consequence of the economics and demographics of industrialization can no longer persist. Indeed, as the nation moves toward zero population growth and a stable percentage of its population in metropolitan areas, growth and development in one set of cities must invariably come at the expense of competing sets. The politics of urban development, therefore, has taken on some of the characteristics of a zero-sum game: For the first time in our history, if there are to be winners in the quest for growth, there must also be losers.

At the same time, the nature of economic growth has altered substantially. Whereas growth before World War II centered on heavy industry and manufacturing, in more recent decades expansion has increasingly involved high technology and service industries. Firms in these fields are less dependent upon access to raw materials and cheap transportation and are consequently relatively unconstrained in locational decision making. The southern rim has benefited disproportionately from this type of economic growth.

Collectively, winter cities have suffered severely in the shift away from heavy industry, experiencing varying degrees of job loss, corporate disinvestment, population decline, and tax-base erosion. As a return to traditional levels of manufacturing is considered unlikely, northern cities are competing vigorously for new types of investment. The new growth industries, however, rely much more than heavy industry on highly educated technical and white-collar workers, workers whose sense of appropriate urban amenities differs substantially from their blue-collar counterparts. In spite of the fact that many of the larger centers, such as Detroit, Cleveland, and Pittsburgh, have long been home to important arts endeavors, they have been hampered in their development efforts by reputations for smoke, grit, and a blue-collar ambience. Creating an environment that will appeal to the new generation of technicians and managers, as well as to the corporate elite who direct the flow of developmental capital, has brought the arts to center stage as a tool in economic development (Barsh and McDonald, 1985; Mason and Skinner, 1981). Support for the arts is in part designed to create an item that city promoters can put in their "plus" columns to balance such negatives as four months of dreary weather and an equally dreary chain-owned monopoly newspaper. That the traditional arts season runs from early fall to mid-spring enhances the arts' appeal: the arts are primarily a winter activity.

Although the arts can be defined broadly to include many types of entertainment such as public television, film, and some forms of music, for the purposes of this chapter we shall confine our discussion mostly to the fine arts: theater, orchestral music, opera, museums, and other permanent arts organizations. These are the arts to which economists have devoted disproportionate attention, and they are also the arts activities that cities most frequently support when they attempt to link the arts with urban development and redevelopment (Hendon, 1985).

The arts contribute to local economies in several ways. First, the arts are themselves an industry, employing workers and producing products for sale. To the extent that arts activity has an appeal beyond the region of production, the arts become an "export" industry by attracting visitors and their expenditures to the producing area. In this manner alone, the arts can have a significant impact on regional economies. Further, through the ripple effect, the arts can foster direct economic development in ancillary areas. Visitors to arts events must be supplied with appropriate accommodations, thus creating markets for hotels, restaurants, and souvenirs. Here the arts function economically in a manner similar to convention centers, commercial spectacles, and nonarts tourist attractions (Chartrand, 1984).

Second, as arts institutions originally clustered in downtown areas and generally have not followed commercial entertainment to the suburbs, urban policymakers have sometimes viewed invigorated arts communities as sources of increased downtown traffic and commercial activity. In this context the arts are expected to function as an "export" industry within the city and to aid in downtown revitalization. Relatedly, works of art or open space for artistic performances are often included in renewal areas to create a more pleasing environment, one in which people will feel comfortable spending leisure time (Fleming, 1981).

Finally, a reputation as an arts center may aid cities in attracting economic development from outside the region. Here support for the arts is an adjunct to larger development strategies. Although a direct link between arts availability and locational decision making is at this point unproven, urban planners often act as if the provision of cultural amenities influences the locational behavior of potential investors. As a result, city boosters "sell" culture in much the same way they "sell" low crime rates, cheap electricity, and docile underpriced work forces (Goetsch and Huderlain, 1983; Shanahan, 1980).

New York City provides a vivid and unique illustration of the potential impact of the arts on a regional economy. A 1980 impact study found that the region's 1900 profit-making and nonprofit arts institutions employed 117,000 people and that arts-related economic activity contributed $5.6 billion to the area's economy. Annually, 13 million people from outside the 17-county New York metropolitan region attend at least one arts event or visit an arts institution. The arts in New York City are big business (Scanlon and Longley, 1984).

But there is only one New York. As the arts capital of the nation its arts-related income is unduplicated. Other cities cannot hope to imitate New York's success, nor is it likely that its arts activity can successfully be lured beyond the bounds of the metropolis. New York and Los Angeles are the major export centers for artistic activity and are almost certain to remain so. It is unlikely that any city in between those two can become a competing arts emporium; modern communications make it too easy for consumers to experience in their homes the highest quality product of the major export centers. That cities that are not now important arts producers stand little chance of becoming so, however, hardly eliminates the arts from a potential role in urban development, although it does suggest that investment and expectations should be along modest lines (Hendon, 1985).

There are a number of obstacles that must be overcome if the arts are to be used successfully as a development tool. First, the arts almost by definition are unprofitable and appeal to a relatively small proportion of the population. Were the case otherwise, what we define as art would in our society be both profitable and a part of mass culture; entrepreneurs would compete to create new sources of supply and discussion would focus on tax revenues collected rather than subsidies expended. The recent failures in both the United States and Canada of several arts cable channels illustrates the problem of insufficient audience for commercial success in the arts generally (Beck, 1985; Thomas, 1984). The segment of the population to which the arts appeal presents a related problem; arts audiences are significantly above average in both education and income. Much recent urban redevelopment has been explicitly designed to attract the white middle class back to the city, elbowing out the nonwhite unwashed who replaced them in the 1950s. Subsidies from limited local resources that benefit almost exclusively the (suburban?) affluent do raise troublesome questions of social equity.

Second, the arts are organized in a manner that makes their inclusion in larger development schema difficult. Virtually every arts institution operates as a separate nonprofit corporation with its own socially prominent trustees, affluent constituency, independent agenda, and carefully nurtured sources of external funding. Cooperation between groups can by no means be assumed, and any urban policy that includes the arts must cope with the institutional fragmentation of the arts community itself. Further, in almost every community any given arts genre is the monopolistic preserve of a single arts institution. There is one symphony orchestra, one ballet company, one art museum; none is likely to share with or encourage the encroachment of competitors. Hence urban development policies that would include the arts are constrained by monopoly-oriented arts institutions that lack a tradition of working together toward larger policy goals (Hendon, 1979: 11; Katz, 1984).

Although local arts institutions are generally monopolies, they tend to be small ones. Except for symphony orchestras and opera, nonprofit performing arts groups seldom employ more than 10 people, and the prospects for growth are not large. Even museum budgets are relatively small. A survey in the mid-1970s found that 60% had annual budgets of less than $50,000 and only 7% exceeded $300,000 (Hendon, 1979: 42). Most creative artists—authors, painters, composers, and photographers—tend to work alone. The organization of work in the arts does not lend itself to traditional firm expansion: a prosperous and busy string quartet is unlikely to respond to increased demand by hiring a fifth player. Hence, large dividends should not be anticipated from policies designed to encourage increased arts employment (Hendon, 1985).

Although arts organizations tend to be small, the arts are nevertheless labor intensive with few prospects for increased productivity. William Baumol and William Bowen have labeled this problem the "cost disease," arguing that as arts productivity will always lag behind increasing industrial productivity, the performing arts in particular will face an ever increasing gap between operating costs and earned revenue. Whereas costs are set by the general cost level in the economy, earned income is limited by the number of performances scheduled, the seating capacity of performing arts halls, and an inability to raise prices as fast as the rise in costs. Thus, maintaining a given level of arts activity without sacrificing production quality or artist income will require ever larger subsidies. From an economic perspective, then, there is little prospect that the arts in general and the performing arts in particular, will in the long run be self-sustaining (Baumol and Bowen, 1968).

In spite of these difficulties the arts have been a popular vehicle for public and philanthropic investment. During the past quarter-century the number of arts centers and performing companies has grown rapidly; indeed, almost 90% of the dance companies performing in 1979 were organized in the previous 10 years. The great expansion in arts-related activity is closely related to changes in funding sources. Since the creation of the National Endowment for the Arts in 1965 public money in relatively large quantities has for the first time been appropriated to underwrite cultural programs. Similarly, states have either created or sharply increased funding for their own arts councils. Public support has provided seed money and in some cases has underwritten operating deficits and has been especially important in bringing the arts to areas lacking in cultural resources. At the same time major corporations have increased their own financial support for the arts, considering it a noncontroversial but highly visible way to sustain illusions of generosity and community service. As arts activities are almost invariably urban, cities have benefited disproportionately from these new sources of support (Lowry, 1984: 133-139; Netzer, 1978: 2, 59-95; Shonberg, 1983).

Attempts in three New Jersey cities illustrate some of the possibilities and problems involved in including support for the arts in urban redevelopment. In New Brunswick, a small city in the central part of the state, the chairman of locally headquartered Johnson & Johnson and an arts supporter led an attempt to revive the city's decaying downtown and develop an arts center. A nonprofit corporation created for the purpose purchased four vacant buildings—a theater, a YMCA, a department store, and a warehouse—as the nucleus for the center. Crucial to the center's success was the participation of Rutgers University, which planned to use one or more of the buildings for its school of art and related activities. Hence, a financially stable arts producer became a major tenant, providing a more secure financial base for the rest of the project (Mori, 1983b).

In Newark, attendance at Symphony Hall fell steadily after the 1967 riot, and the hall closed in 1976. Metropolitan Opera singer Jerome Hines became the spark behind the Save Symphony Hall Committee, which with foundation and government grants reopened the hall a year later. But perception of the area as unsafe remained, and in 1983 the committee planned as soon as possible to add a 527-car garage and a restaurant. Suburban patrons would then be able to arrive and leave without setting foot on the streets of Newark (Mori, 1983a).

In Rutherford, a suburb in the shadow of New York, the William Carlos Williams Center for the Performing Arts was an imaginative attempt to renovate a historic theater. An ambitious project, the center depended on two movie theaters completed early in the renovation to provide the revenue for debt service and further construction. But the theaters were less successful than anticipated, and in 1983 the center entered default, its future in doubt (Mori, 1983c).

Although proposals to support the arts with public funds tend to be relatively noncontroversial, public response to a plan for a performing arts center in Stamford, Connecticut, in 1982 illustrates the type of opposition that can develop. To fund the $5 million project, the state government allocated $1.5 million, and local corporations pledged an additional $3 million. Full funding depended upon $500,000 to be raised by the city through the sale of municipal bonds. City financial support for the center, however, became a focus for taxpayer discontent. Opponents, an anonymous city official told the *New York Times*, saw the center as "a leisure activity for the rich and nonresidents, as well as an unpleasant reminder of the encroachment of downtown development on their lifestyle." Supporters, on the other hand, considered "the undertaking crucial to the development of the city's prestige and stability. They regard the center as the gemstone of the downtown renewal area." The city's Board of Representatives approved the bond issue by a margin of just one vote (Charles, 1982; "Connecticut Journal," 1982).

Elements of prestige and urban rivalry frequently appear among the reasons for using the arts in urban development. Urban elites have worked to increase cultural offerings for the same reasons they have worked to attract professional sports franchises: to show that their cities are truly "major league."[1] The Dallas Arts District is a carefully planned 60-acre expanse at the north end of downtown devoted to museums and the performing arts. But Dallas novelist A. C. Greene claims that although the city supports the existence of artistic institutions it is less committed to what those institutions do. Dallas, claims Greene, "supports symphony, opera, ballet—but not composers, singers, choreographers." Dallas is not unique; almost everywhere arts facilities have expanded faster than repetoire. Programming, especially in smaller places, tends to be conservative and to repeat the familiar in hopes of attracting the largest possible audiences. Few institutions—the Walker Arts Center in Minneapolis and the Contemporary Arts Center in New Orleans are notable exceptions—regularly schedule avante garde or experimental fare (DiMaggio, 1984; Freedman, 1984; Goetsch and Haderlain, 1983; Shonberg, 1983).

Interest in the arts as an urban development tool, especially in winter cities, has grown as interest in the arts themselves have grown. The arts have always appealed to the educated affluent, an expanding population category in modern American society. Further, it is to these people and to the "high-tech" industries that employ increasing numbers of them that cities are turning as a source of future growth. Hence, the arts become part of a sort of "amenity infrastructure" that, it is hoped, will translate not only into an urban reputation for those qualities thought to appeal to professionals and executives, but also into capital investment that might otherwise go elsewhere. Yet the arts themselves produce but modest levels of employment and seldom produce enough revenue to cover costs. Often "support for the arts" is more vigorous for buildings and organizations than it is for the content of artistic experiences. For the arts to succeed in a locality there must be a core of individual and institutional patrons sufficiently committed to the arts for their own sake to maintain interest and underwrite deficits from season to season. Only then can the arts make a regular and sustained contribution to the winter life of winter cities.

NOTE

1. Peter Zeisler, past chairman of the National Endowment for the Arts' theater panel, summed up the relationship between the arts, professional sports, and urban image at a symposium in New York in 1983:

> Half the people that contributed to some of the theatres being built around the country didn't give a good damn about the theatre, but they knew that in order to be a main line city, to be an important city, they had to have cultural amenities. You know, it really is related to the spread of professional sports in this country. When a city got a major league football team or a major league baseball team, then it had to have a full-time symphony orchestra, it had to have a major repertory theatre, it had to have a ballet company [Lowry, 1984: 113].

REFERENCES

BARSH, F. and A. McDONALD (1985) "Supply of performing arts: financing a cultural center," in V. L. Owen and W. S. Hendon (eds.) Managerial Economics for the Arts. Akron: Association for Cultural Economics.

BAUMOL, W. and W. BOWEN (1968) The Performing Arts: The Economic Dilemma. Cambridge, MA: MIT Press.

BECK, K. (1985) "Cultivating the wasteland: the U.S. cultural cable experience," pp. 67–81, in M. A. Hendon et al. (eds.) Bach and the Box: The Impact of Television on the Live Arts. Akron: Association for Cultural Economics.

CHARLES, E. (1982) "Art center in Stanford is uncertain." New York Times (May 9): Sec. 23, p. 1.

CHARTRAND, H. (1984) "An economic impact assessment of the Canadian fine arts," in W. S. Hendon et al. (eds.) The Economics of Cultural Industries. Akron: Association for Cultural Economics.

"Connecticut Journal" (1982) New York Times (May 23): Sec. 23, pp. 1.

CWI, D. (1985) "Changes in the U.S. audience for the arts," pp. 32–42 in C. Richard Waits et al. (eds.) Governments and Culture. Akron: Association for Cultural Economics.

Di MAGGIO, P. J. (1984) "The nonprofit instrument and the influence of the marketplace on policies in the arts," pp. 57-99 in W. McNeil Lowry (ed.) The Arts and Public Policy in the United States. Englewood Cliffs, NJ: Prentice-Hall.

FLEMING, R. L. (1981) "Recapturing history: a plan for gritty cities." Landscape 25: 20–27.

FREEDMAN, S. G. (1984) "Reporter's notebook: for Dallas a growing cultural identity crisis." New York Times (August 6): sec. 3, pp. 17.

GOETSCH, R. and M. HADERLAIN (1983) "Art for downtown's sake." Planning 49 (July/August): 10–14.

HENDON, W. S. (1979) Analyzing an Art Museum. New York: Praeger.

——— (1985) "Arts and the economic life of the city." Poetics.

KATZ, S. N. (1984) "Influences on public policies in the United States," pp. 23–37 in W. M. Lowry (ed.) The Arts and Public Policy in the United States. Englewood Cliffs, NJ: Prentice-Hall.

LOWRY, W. M. [ed.] (1984) "A symposium: issues in the emergence of public policy." in The Arts and Public Policy the United States. Englewood Cliffs, NJ: Prentice-Hall.

MASON, P. F. and D. J. SKINNER (1981) "Raking in the chips." Planning 47 (November): 10–13.

MORI, R. (1983a) "Performing arts growing in state at a high cost." New York Times (April 3) sec. 11, pp. 1–2.

——— (1983b) "Arts centers struggle to keep their audiences and books." New York Times (April 10): sec. 11, p. 2.

——— (1983c) "Williams Center teeters on brink." New York Times (April 17): sec. 11, pp. 8–9.

NETZER, D. (1978) The Subsidized Muse: Public Support for the Arts in the United States. New York: Cambridge University Press.

SCANLON, R. and R. LONGLEY (1984) "The arts as an industry: their importance to the New York-New Jersey metropolitan region," pp. 93–100 in W. S. Hendon et al. (eds.) The Economics of Cultural Industries. Akron, OH: Association for Cultural Economics.

SHANAHAN, J. L. (1980) "The arts and urban development," pp. 295–305 in W. S. Hendon et al. (eds.) Economic Policy for the Arts. Cambridge, MA: Abt.

SHONBERG, H. C. (1983) "Have cultural centers benefitted the arts?" New York Times (July 10): sec. 2, pp. 1, 26.

THOMAS, T. J. (1984) "The failure of culture on pay television," in W. S. Hendon et al. (eds.) The Economics of Cultural Industries. Akron, OH: Association for Cultural Economics.

13

Food Policies for Cities in Northern Latitudes

JON VAN TIL and MEDARD GABEL

□AS IS THE case with so many other urban problems—energy, resource recovery, and water quality among them—we are beginning to rethink the problem of providing food to urban residents. Long seen as simply a matter of transportation, distribution, and marketing, the problem of urban food may be seen to involve questions of nutrition, production, and employment as well.

Contemporary work in a number of locations—the Center for Neighborhood Technology in Chicago, the Graduate School of Planning at the University of Tennessee, the Cornucopia Project of the Rodale Press, and the Program on Developing Municipal Food Policies of the U.S. Conference of Mayors—is reviewed in this chapter. From this work emerges the outline of a new food policy for winter cities.

We begin with questions of who, what, when, where, and why that require answers if a new food policy is to be developed: *Who* gets the food? *What* is the food? *When* food isn't there, what happens? *Where* is food grown? *Why* is recasting the problem important?

First, who gets the food? This distributional question, long thought banished as a problem requiring domestic concern, has resurfaced in recent years. With hundreds of thousands of individuals homeless in our cities (U.S. Department of Housing and Urban Development, 1984), over 20 million persons receiving food stamps nationally, and a problem of perennial underemployment (Judis and Block, 1985), the U.S. city has come to see hunger as not just a problem of the Third

---

AUTHORS' NOTE: *We are grateful for the criticism of an earlier draft of this chapter by Professor Roy Van Til of the Economics Department of the University of Maine at Farmington.*

World. We too need to worry about getting enough food to all our citizens; we too need to worry about hunger and starvation at home. Although this question most directly reflects decisions about the distribution of income, it may be affected as well by how and where we grow our food.

Second, what is the food? We Americans recognize our accustomed diet of meats, carbohydrates, vegetables, and sweets. But do enough of us realize that other diets might serve our bodies at least as well at a far smaller cost? For every calorie of food we produce, 6.4 calories of energy have been expended (Schwartz-Nobel, 1981: 115). For every pound of beef a cow gains in its feedlot, between 7 and 16 pounds of grain have been consumed (Ford, 1978: 194). Cheaper soybean, cereal, and alternative protein-based diets have been shown to reduce the dollar and health costs of raising and consuming fat-laden edible meats (Ford, 1978; Lappé, 1971).

The third question involves the possibility that food may run short in the face of even a limited state of emergency in the winter city. A typical U.S. city holds no more than a three-day supply of food. Stores typically hold most of this reserve on their shelves with only limited additional food in storage. These stocks can disappear in hours in a food shortage or panic-buying spree. The interruption of normal supply patterns by a severe weather emergency, ecological disasters like the Medfly outbreak in California, labor-management disputes like the most recent truckers' strike, nuclear power-plant accidents, or threat of war would suffice to leave most stores and many homes short on food within a matter of hours.

Fourth, the problem of where food is grown has entered contemporary discussions. The Cornucopia Project of Rodale Press has pointed out that U.S. food supply lines are overextended—that the average molecule of food travels over 1300 miles from point of production to point of consumption. They advocate the development of more regional markets by winter cities, seeing gains in food freshness, nutritional value, taste, transportation costs, economic security, and local economic development.

Finally, the overall significance of the problem of urban food has increased in light of our growing international awareness of the fragility of food systems. The tragedy of the contemporary African food crisis is shadowed by the awareness that to some lesser but significant extent it could also happen here. Schwartz-Nobel puts it this way (1981: 27):

> Hunger is not simply the result of scarcity or poverty. Energy shortages, inflation, the destruction of vital farmland, the failure of crops to respond to chemicals, problems of food distribution, and the mismanagement of food production all affect food supply and increase food prices. Evidence of an emerging pattern of dislocation is everywhere.

Without the massive food assistance programs in place in the United States (food stamps alone cost $11.6 billion in 1984), many of the 34 million people who live below the U.S. government-defined poverty line would be in danger of starvation. With these programs, many are still malnourished. Famine is at least as much a function or result of lack of employment as it is of gross food production. As the head of the International Rice Research Institute in the Philippines has pointed out, "The world suffers from a famine of jobs, not a lack of food production."

FOOD SYSTEMS IN CITIES

Urban food systems have been described and analyzed in the literature. In this section we review the nature of these systems, explore their strengths and weaknesses, and consider their possible futures.

Paige Chapel describes the changing Chicago food system (1982: Part I,1):

> Many years have come and gone, as well as farmers, manufacturers and retailers, since Chicago was hog butcher to the world. Each year, the Chicago area loses land that was devoted to the production of food; fewer Chicagoans are employed by the food industry as the number of warehouses and factories leaving Chicago continues to grow; and more and more of the retail food market becomes concentrated in fewer corporate headquarters. The final result is a highly complex food chain that depends upon an intricate transportation network, distant resources, and a sophisticated market economy—a food chain that few consumers understand or have access to.

Working under the direction of Robert Wilson, a team of graduate planning students at the University of Tennessee describe the "Knoxville urban food system" as "large, diverse, and difficult to recognize as a system. There seems to be little cooperation at local levels of management" (Blakey et al., 1982: 2). Despite its nebulousness, the students identified "six components of the urban food system: production components, manufacturing and processing components, wholesale distribution components, retail distribution components, 'preparation for consumption' components, and consumption components" (Blakey et al., 1982: 2–3).

When these food systems are analyzed, as they have been in Chicago and Knoxville, a number of special concerns emerge. In Knoxville they were identified by Wilson and his students as the loss of agricultural land and near-Knoxville food production, the lack of wholesale produce facilities, urban food transport, supplying the 1982 Expo, and food assistance for the disadvantaged (Blakey et al., 1982: 101). In Chicago, the overriding concern was identified as the lack of available and affordable food to a half-million of the city's residents (Chapel, 1982: Part II,3).

To Chapel, Chicago's problems appeared relatively intractable: "Much of the difficulty in correcting what is wrong with the food chain is that the system currently seems so right in so many respects" (1982: Part II,18). Many food experts marvel at the sophistication and complexity of the system, and many economists point to its ability to maximize the food desired by rational consumers for the cheapest price possible. These defenders of the market ask: Why not employ land for its highest economic use? Why not buy at the large supermarket where prices are cheaper than at the "mom and pop" store on the corner? Why not continue to believe that the customer is right and not manipulated by advertising when he or she continues to eat fast-food beef rather than tofu? Why not trust our food consumption patterns to the institutionalized economies of corporate and global interdependence?

Economists Gar Alperovitz and Jeff Faux (1984: 202) explain why the food system does not work as well as classical economic theory dictates:

> The increasingly concentrated and centralized structure of corporate control in food processing and distribution contributes to rising prices all down the line. . . . Concentration contributes to excessive food prices through higher profits, managerial salaries, and advertising expenses than would prevail in competitive markets.

Alperovitz and Faux call for the prevention of monopoly control in food markets and for the expansion of "smaller human-scale institutions." They note that over the past several decades, a "rich variety" of local experiments in such states as Pennsylvania and Massachusetts have contributed to development. "These institutions include various forms of 'direct' marketing, including selling by farmers directly to consumers at farmers' markets; through nonprofit food stores; and to institutions such as schools and hospitals. Alternative marketing efforts also include consumer food cooperatives and food-buying clubs." Food coops, they note, offer savings that range up to 40% (Alperovitz and Faux, 1984: 203).

The Knoxville report presents a set of 15 recommendations to achieve human-scale food institutions. Their list includes the establishment of a regional food council, the preservation of agriculturally productive lands in the region, the development of urban gardening, the improvement of the local food circulation system, the coordination of food delivery, the development of disaster awareness, and the development of programs to feed the disadvantaged adequately (Blakey et al., 1982: chap. 5). The report concludes with a call for the development of a vigorous public-private partnership dedicated to the improvement of this "vital urban life-support system" (Blakey et al., 1982: 159).

As Alperovitz and Faux note (1984: 203), we will have to move toward "greater regional self-sufficiency in the production of certain commodities" to save transportation and energy:

> We now truck vegetables an average of two thousand miles to market! State programs to rebuild truck gardening have had considerable success recently. We can also further this goal through the use of vacant urban land for cooperative community gardens. With better planning and the use of organic approaches and appropriate technology, food production as a whole could be a net producer of energy for the rest of society, rather than the large consumer it now is. More local production combined with the restoration of healthy, locally based processing firms such as bakeries and canneries contribute not only to stable and lower food prices but also to stable communities.

The literature reviewed thus far suggests that awareness is unlikely to develop suddenly regarding either the importance of, or the necessity to modify, the urban food system. Yet it is possible that the food system's weaknesses could become glaringly apparent in a matter of mere hours in any urban area.

To aid in the consideration of possible futures of the urban food system, we employ the four major societal scenarios elaborated by Van Til (1982): the two "good luck" scenarios of "Reagan's hope" and "transformation," the continuity scenario ("muddling through"), and the "hard luck" scenario of shortfall and suffering.

The first image of the future, "Reagan's hope," envisages a future of increasing affluence and technological plenitude. In this image urban food systems will increasingly flourish by means of their conventional reliance on national (and even international) production, chemical preservation, intricate transportation, imaginative marketing, and consumer "sovereignty." Hunger will be banished as an epiphenomenon, and will disappear along with poverty into the oblivion of either history or the realm of the unnoticed.

The second image, "societal transformation," involves a widespread acceptance of new values of ecological awareness and human concern. In this future the food system will become a matter of broad citizen awareness. Diets will shift rapidly away from meat toward other forms of vegetable protein; gardening and small-scale agriculture will flourish; and both poverty and hunger will be eliminated in an economy of mutual concern and care.

The third image, "muddling through," assumes that the future will be much like the past. Occasional crises will arise, but the system will adapt, incrementally and effectively. Salad bars will increase in number and popularity; the consumption of white wine and juices will continue to rise as that of whiskey, beer, and milk decreases; and gardening will be a somewhat more popular urban and suburban hobby. But the urban food system will remain essentially an unknown operation, and any changes will remain a matter of producer preference.

Finally, the shortfall scenario envisages a world ensnared by resource scarcity—energy, food, water, land, air, and capital. In such a scenario prolonged and recurrent food shortages will plague our winter cities. Diets of even the middle class will suffer, and many lower-class persons will die of malnutrition each year. Land and topsoil will become barren in the wake of ever growing chemical pollution and mineral exhaustion. The United States will join much of the rest of the world in the vise of famine and poverty.

Which of these scenarios will prevail? Surely we cannot predict, for all are possible. Which should prevail? Which insulates us best from the potential ravages of fate and folly? We will show that we do have a choice. A future of increasing food consciousness and self-reliance is described as the path that best lets us hope for the best while assuring against the worst.

COPRODUCING FOOD: WHAT IS POSSIBLE?

Contemporary observers of the urban service economy have increasingly noted the presence of "coproduced" services—services whose quality and quantity is dependent upon efforts of both consumers and service providers (Rich, 1981). The provision of food, even to urbanites, seems no exception to this pattern.

Granted, it is conventional for the urbanite to see food as something that is grown elsewhere (on the farm), processed in some middle location, and then trucked to the city store for purchase and consumption. Yet, some will recall that during World War II, Americans grew in over 20 million Victory Gardens some 40% of the vegetables consumed. Today, there are over 42 million gardens in the United States, producing about $300–$500 worth of food apiece, and amounting to over $12.5 billion worth of food per year. Contemporary urban gardens, though providing a small part of this staggering total, have provided many urban families a significant portion of their vegetable intake throughout American history (see Russell, 1976).

In various cities throughout the world, remarkable productivity can be seen, as Robert Rodale (1984) has observed:

> People in some cities are remarkably efficient at producing their own food. Shanghai residents, for example, grow all their own fresh vegetables. And even Hong Kong, with little spare land, produces 42 percent of its green vegetables. In Brazil, the rights-of-way under high tension lines in some cities have been converted to gardens. Animal production in cities is rising too. Singapore produces 80 per cent of its chickens and much pork. Small livestock of several kinds thrive in Latin American towns.[1]

Urban gardening in European towns and cities is a long-followed tradition. In back yards and plots surrounding towns, vegetables are grown to supplement store-bought diets. The food contribution of European urban gardening, though never fully studied, is suggested by Ignacy Sachs's observation that Paris in the nineteenth century was a net exporter of lettuce (personal communication, 1984). Using composted horse manure and hothouses, this production more than met the city's ravenous demand for salad consumption. Similar efforts to capture the heat emitted by power plants are providing the energy for hothouses in contemporary Pennsylvania.

Farming on the urban fringe, as Roger Blobaum (1984) notes, can provide for agricultural stability, increased off-season production, direct marketing, and waste recycling. A "Metropolitan Food Plan," Blobaum notes (1984: 57), could "strengthen the connections between the economic viability of close-in farms, the preservation of farmland in the urban fringe, and the food needs of urban consumers." Large tracts of undeveloped land are available in most cities, and the problem of preserving remaining exurban land from corporate and residential sprawl is an important one. A metropolitan agricultural policy, enforced by property tax classification and active use of zoning laws, might assist in retarding the seemingly endless encroachment of suburbanization in major metropolitan corridors.

Once the urbanite begins to see food as something that she or he can participate in producing, a new awareness of the quality and use of food might be expected to develop. Such an emerging consciousness is reflected in the dramatic shifts in the dietary preference of some Americans in recent years. The movement toward lower fat intake, reduced red meat consumption, and an awareness of food content, though not yet characteristic of most Americans, has certainly taken hold of a significant minority. Evidenced by the spread of health food, salad bars, and nutrition magazines, the movement toward coproduced food use may well by one that will expand considerably in the years ahead.

URBAN FOOD POLICY AS AN EMERGING FIELD

Food policy, the U.S. Conference of Mayors (1984) has noted, "has rarely been included in a city's comprehensive planning activities." And yet the relationship between food supply and other elements of any city's public policy, "such as transportation, economic development, housing, health and sanitation, regulatory measures, and emergency preparedness," is clearly demonstrable. The Conference of Mayors has recently undertaken to establish a program to encourage cities to develop food policies. Four cities—Charleston, Kansas City, Philadelphia, and St. Paul—have been identified to receive special assistance from this program.

The work of this project allows us to see at least the outlines of a developing food policy for American cities. Some elements of this policy are emerging in a context of wide consensus; others are more highly controversial, and give rise to vastly different orientations.

Among the elements that have initially found broad agreement are the following:

(1) Food supply should no longer be taken for granted.
(2) Food production, processing, and distribution employ many people in the urban environment and could employ more with the right mix of incentives.
(3) Food assistance for the needy requires skillful coordination.

Among the elements that have aroused initial controversy are the following:

(1) Local food production can have a significant impact on overall food supply.
(2) The use of local vacant land for food production instead of development should be considered.
(3) Cities would be well advised to establish their own Department of Food to assure a continuous and nutritious supply of food.

As this policy discussion has developed, clear distinctions have begun to emerge between positions that are essentially centralist and those that are fundamentally decentralist. The centralist position emphasizes the specialization of agricultural development and the efficiencies of processing and transportation technologies. The decentralist position, on the other hand, focuses on the perceived need to shorten lines of production and distribution.

One's choice between these two positions, or the creation of an intermediate one, depends to great extent upon one's confidence in the ability of the transportation-distribution system to withstand the kinds of disruptions occasioned by weather, labor-management tension, or environmental crisis. In this area as others, the optimal fair-weather system may prove inadequate in a time of crisis.

The decentralists, among whose number we count ourselves, argue that although it is wise to hope for the best, it is prudent to plan for the worst. And there are values to be achieved from decentralization that elude the centralists: diet consciousness, improved regional economics, regional pride, and food security among them.

From our perspective, a city concerned with its food system would be wise to take the following kinds of initiatives:

- Establish a Department of Food that would encourage local food production and distribution through its policies and planning for the meeting of local needs in both normal and emergency times.

- Preserve land by establishing land-use management committees, facilitating the establishment of urban block farms, and creating agricultural open-space preserves.
- Develop and encourage local food production by making selected publicly owned lands available for urban gardens, providing local tax incentives for home/community produced food, establishing Urban Agricultural testing programs, subsidizing gardens, encouraging the use of waste heat for greenhouse support, and establishing large-scale composting operations.
- Develop and encourage local markets by assisting in the establishment of farmers' markets in the city, assisting in the formation of buying clubs and food cooperatives, and providing incentives for supermarkets to remain in inner cities.
- Develop and encourage local food processing and storage by sponsoring community canning facilities and providing incentives to grocery stores to provide such facilities.
- Facilitate access to food by assuring that the needy are receiving required food services, providing transportation to farmers' markets, and establishing a "food and hunger hotline."
- Educate people about their food situation through public meetings and appropriate school programs.

The policy discussion presently taking shape gives promise of being a productive one, and there is evidence that persons of different views are listening carefully to each other. In the balance of the discussion ride crucial interests of the winter city, whose food policy will surely continue to emerge as a necessary element to its survival in the twenty-first century. The participation of urbanists and urban residents alike in this process will surely help shape an outcome assuring that urbanites will enhance their provision of that very basic necessity, food in the city.

NOTE

1. Documentation for these claims is found in *Mazingira* (1984).

REFERENCES

ALPEROVITZ, G. and J. FAUX (1984) Rebuilding America. New York: Pantheon.

BLAKEY, R. C. et. al. (1982) Food Distribution and Consumption in Knoxville: Exploring Food-related Local Planning Issues. Emmaus, PA: The Cornucopia Project of Rodale Press. (reprinted)

BLOBAUM, R. J. (1984) "Farming on the urban fringe," pp. 55–61 in F. R. Steiner and J. E. Theilacker (eds) Protecting Farmlands. Westport, CT: AVI Publishing.

CHAPEL, P. (1982) The Chicago Food System: Research Findings of a Study in Two Parts. Emmaus, PA: The Cornucopia Project of Rodale Press. (reprinted)
FORD, B. (1978) Future Food: Alternative Protein for the Year 2000. New York: William Morrow.
JUDIS, J. and F. BLOCK (1985) "No more jobs." Not Man Apart (Friends of the Earth): 12–13.
LAPPÉ, F. M. (1971) Diet for a Small Planet. New York: Friends of the Earth/ Ballantine.
Mazingira (1984) "Urban gardens may feed the urban poor." (July): 16.
RICH, R. C. (1981) "Municipal service and the interaction of the voluntary and government sectors." Administration and Society 13: 59–76.
RODALE, R. (1984) "Cities can sprout food." Speech, Emmaus, PA.
RUSSELL, H. S. (I976) A Long, Deep Furrow: Three Centuries of Farming in New England. Hanover, NH: University Press of New England.
SCHWARTZ-NOBEL, L. (1981) Starving in the Shadow of Plenty. New York: McGraw-Hill.
U. S. Conference of Mayors (1984) Developing Municipal Food Policies. Washington, DC: Author.
U. S. Department of Housing and Urban Development, Office of Policy Development and Research (1984) A Report to the Secretary on the Homeless and Emergency Shelters. Washington, DC: Author.
VAN TIL, J. (1982) Living with Energy Shortfall: A Future for American Towns and Cities. Boulder, CO: Westview.

Part III

The Revival of Industrial Cities

□THE LIVABILITY of winter cities is an important and necessary but insufficient condition for their revival. Their economic revitalization also requires a vigorous reshaping of the local environment for business development.

In an effort supported by the George Gund Foundation, a number of corporate chief executive officers in Cleveland organized in the early 1980s an effort called "Cleveland Tomorrow." It produced "a strategy for economic vitality" that briefly stated that Cleveland's ultimate success lay in a four-pronged approach consisting of these objectives:

- retain the existing manufacturing base through an improvement of labor productivity and quality of work life in local plants;
- foster new business development through both entrepreneurship and advanced industrial technology;
- assist in the physical rebuilding of the central city through better planning, organization,and development of catalytic projects; and
- assure essential public support for education, economic development. public financing, and infrastructure.

Although formulated for Cleveland (Shatten, 1985), these four objectives are probably generic for any winter city that has traditionally been dependent upon manufacturing for its principle employment base. Each of the four objectives can be discussed in more detail to elaborate their generic value for other cities.

First, it is important to note that the decline in manufacturing jobs is not going to be a decline to zero jobs. It is true that as a share of total employment, manufacturing employment has fallen from approximately 35% of total U.S. employment in 1953 to about 20% in 1985.

But in an economy with a labor force of over 100 million, that still means 20 million manufacturing employees. Through a combination of union "givebacks" and productivity improvements, an urban labor force can compete to be part of the remaining industrial labor force.

Second, even with the huge deficit and overvalued dollar, manufacturing output is up. From 1970 to the mid-1980s the index of industrial production has risen about 50% from 106 to 158 (1967 = 100). This increase in industrial productivity is based upon both new and better technologies and the growth of newer and smaller firms. Both industrial research and venture capital have become new public policy issues in cities and will remain so (Johnson, 1984), especially in the older industrial cities.

Third, the restructuring of urban form in the older industrial cities requires both a dramatic vision and an expansion of the scale and quality of planning. It took imagination and boldness to convert the original Quaker Oats factory and silos in Akron into a shopping center and hotel convention center. Similar vision was required to convert the old Sherman tank plant on the outskirts of the Cleveland airport into an international exposition center. Most older northern cities can point to similar examples and to other proposed efforts that are retarded by zoning constraints of a previous era.

Fourth, as Robert Hamrin (1980) and others point out, a new economics is emerging and the new economy requires a new approach to education. The norms of industrial schooling are no longer appropriate in an economic system that is driven by information and innovation (Gappert, 1979).

SECTION OUTLINE

In this section Thompson and Thompson elaborate five distinct paths to local economic development. Although their analysis is specific to the greater Akron-Cleveland-Elyria metropolitan area, their generic framework is appropriate for any urban industrial region.The five development paths include the following:

- routine operations in manufacturing, trade, and services;
- precision operations requiring higher skill levels;
- research and development in science and technology;
- central corporate administration; and
- entrepreneurships and the development of new businesses.

These five paths are described as "sometimes substitutive, partly complementary and often sequential."

Using data from over 30 urban regions or metropolitan areas, their analysis of comparative advantage supports the concepts of Jane Jacobs with regard to the economic significance of ingenuity and inventiveness as a mainspring or urban vitality and resurgence. Although their analysis shows that the industrial Midwest is below average on almost all measures of entrepreneurship, this disadvantageous condition is susceptible to intervention at both an individual and institutional level.

In the next chapter Hanson develops a profile of urban development in an advanced industrial economy. He argues that a new urban system has emerged, which is characterized by a few large and diversified "command and control centers" and a large number of "subordinate centers." The economics of the command and control centers are shifting dramatically from a manufacturing to a service base while the subordinate centers are becoming both more specialized and more dependent upon decisions made elsewhere.

Hanson indicates that cities are in the grip of a massive and pervasive economic and social transformation as important to their future as was the industrial revolution of the late nineteenth century. But from a strategic perspective the new revolution can be foreseen and understood. Drawing upon his earlier work with the Committee on National Urban Policy, Hanson indicates that urban development policies and strategies need to deal simultaneously with both the mainstream of change and those left behind. He cites several northern cities that have developed policies supporting their assumption of new roles in the advanced economy that are quite different from those they played in the older industrial era.

For the most part Hanson considers that Snow Belt cities continue to have a considerable competitive advantage over most Sun Belt cities in a number of factors essential to growth in an advanced industrial economy including (a) skilled labor, (b) well-developed infrastructure, (c) good education and other public services, and (d) strong cultural and other nonprofit institutions.

Buss and Redburn discuss the politics of revitalization and closely examine the role of public subsides to support private initiatives. Although they conclude that some direct subsidies, used very selectively, may support the right kinds of private investment, Buss and Redburn indicate that they are not the key ingredient in successful local development. They agree that the primary public sector's responsibility should be with appropriate infrastructural provision and institutional development.

In their chapter Blair and Dung attempt to develop a framework for the coordination of both proposed industrial and urban policies with macro economic policies. Their simulations suggest that none of the three policy strategies would be sufficient without being supplemented by relocation efforts that would move (or motivate the movement of) people to the places where job growth was above average.

The idea of relocation assistance was first formulated in the report of the President's Commission for a National Agenda for the Eighties, which was released in the last weeks of the Carter administration. Like the idea of "planned shrinkage" advanced earlier by Roger Starr, it received a reaction of denial and rejection. But the analysis provided by Blair and Dung should lead to the reappraisal and further formulation of the concept of relocation policy. Its special significance for winter cities especially as it affects the elderly and the older worker should not be neglected.

An examination of the revitalization, or lack thereof, of new England cities is provided by Howell. The long, slow, but sustained recovery of New England is by no means uniform. Howell's analysis, avoiding any sweeping generalizations, provides a thoughtful and realistic appraisal of New England's urban conditions that is certain to be applicable elsewhere.

Finally, analysis of the transformation of Akron from a blue-collar manufacturing city to the home off our major multinational corporations is provided by Costa, Dustin, and Shanahan. Their analysis of Akron's plans to continue its redevelopment as a high-tech polymar valley also emphasizes the physical reforming of portions of the city to better incorporate its new and evolving functions.

Their analysis serves to reinforce the assumption that cities in northern latitudes, especially the older manufacturing cities, will look as different in the future as will be the functions they serve and the employment they provide.

REFERENCES

GAPPERT, G. (1975) Post-Affluent America. New York: Franklin Watts.

HAMRIN, R. D. (1985) Managing Growth in the 1980's: Toward A New Economics. New York: Praeger.

JOHNSON, C. (1984) The Industrial Policy Debate. San Francisco: Institute for Contemporary Studies.

SHATTEN, R. A. (1985) "Public forum: starting new companies in Cleveland." REI Review 2 (June).

14

Alternative Paths to the Revival of Industrial Cities

WILBUR R. THOMPSON and PHILIP R. THOMPSON

□THE PROLIFERATION OF attempts to provide a relative measure of state business climates was a major impetus to the development of the alternative, five-path, occupational-functional approach to local economic development planning offered in this chapter. The temerity of giving the various states grades for locational attractiveness was bound to spark controversy—ranging to outrage—among hypersensitive area-development groups, especially considering the criteria chosen for inclusion and the weights given them in the indexes. But whatever grade one might be inclined to give the graders, these first efforts at quantifying comparative advantage in business location did provide the needed impetus to inspire further efforts such as this one.

Two basic criticisms of these state business climate indexes are that (1) they designate a whole state as a local economy in business locational decisions, thereby aggregating many separate local labor markets with very different characteristics into a single, bland statewide average, and (2) virtually all of the dozen or more indexes used favor the lowest-cost place in which to perform simple operations and penalize places that have evolved to perform sophisticated manufacturing or commercial operations.

The very process of local economic development erodes the standing of the community in any composite index dominated by average wages and taxes. Economic progress raises skills and wages, increasing, in turn, income and the concomitant demand for local public services, which acts in turn to push up local tax rates. One needs an alternative scale on which to score mature economies that have become overqualified for routine work.

Our dissatisfaction with this implicit index of the best location for routine operations coincided with a contemporary effort on our part to construct a scale on which to measure the technological climate of a metropolitan area (REI Review, 1983). Since that time our work has been extended to encompass the following five distinct paths to local economic development along which to track the position and prospects of a local economy:

(1) routine operations, ranging across manufacturing, trade, and services;
(2) precision operations focusing on industries that employ more artisans and technicians;
(3) research and development, serving as a center of science and technology;
(4) central administration, performing the headquarters function; and
(5) entrepreneurship, capturing the propensity to start new businesses.

To avoid the shortcoming of working with an overaggregated state economy, these data were collected by metropolitan area.

Finally, these five paths are sometimes substitutive, partly complementary, and often sequential in the long process of local economic development. And this occupational-functional approach is itself complementary to the traditional industry-targeting strategy.

AN OCCUPATIONAL-FUNCTIONAL APPROACH

In reading the description of the five paths to local economic development below, which uses about three dozen variables, it may be useful to refer to the figure depicting the profile of the Cleveland metropolitan area. This figure is taken from the Appendix.

POSITIONS AND PROSPECTS IN ROUTINE OPERATIONS

Local economic development strategists in the mature industrial areas must get around the high local wages and fringe benefits for routine work, a product of coupled oligopoly and union power. They repel not only competitive industries but also rollout from the export industries into the local service sector, including municipal workers, raising the cost of living and the local taxes. This acts to prompt new rounds of wage demands and further increases the local cost of doing business.

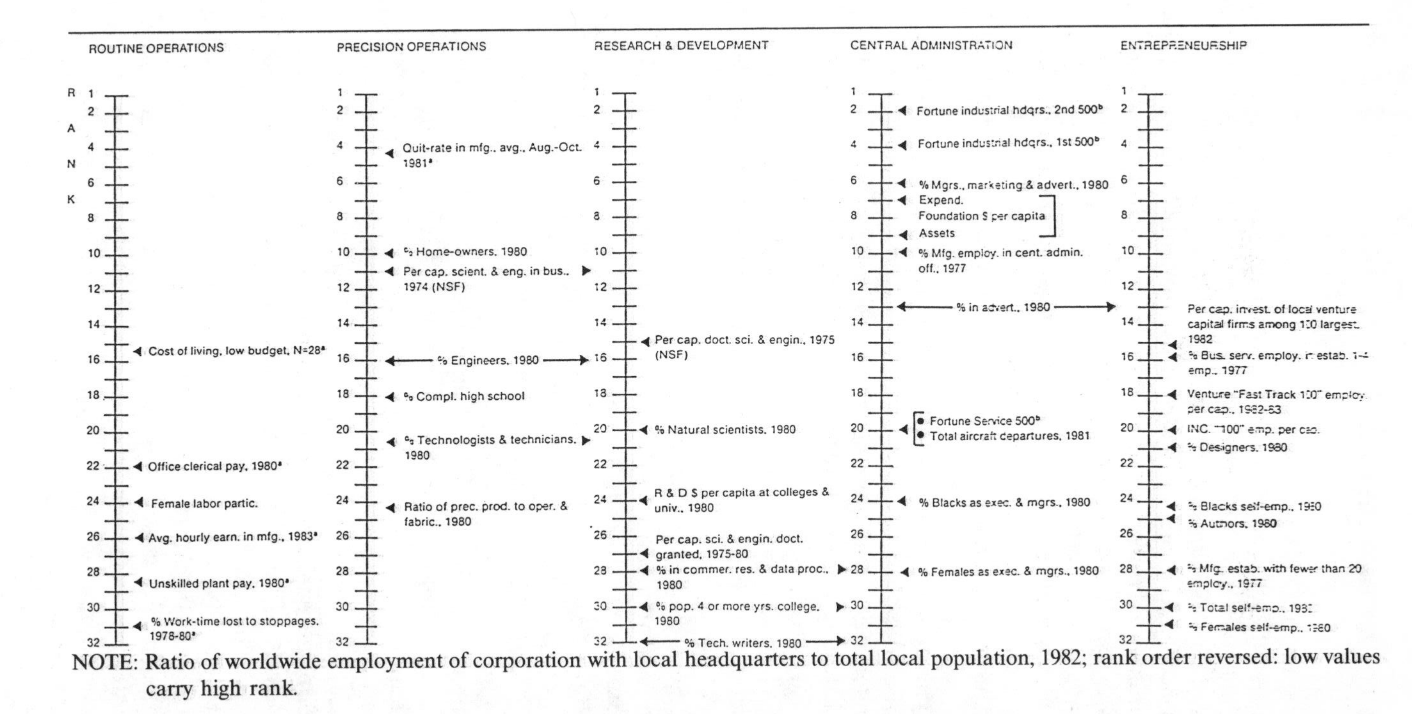

NOTE: Ratio of worldwide employment of corporation with local headquarters to total local population, 1982; rank order reversed: low values carry high rank.

FIGURE 14.1 Comparative Position of the Cleveland-Akron-Lorain-Elyria Consolidated Metropolitan Area Among the 32 Largest Areas on Selected Measures Along Five Alternatives Complementary/Sequential Economic Development Paths

In the long run, this easy money constrains local progress in education by enticing students to dropout of high school and college. Note that high average hourly earnings that reflect high skills promote local economic development through richer local markets, better public services, and a higher-quality environment in almost every way. Unionization typically is uninviting to routine business operations, and is seen as being likely to cause higher labor costs, more work stoppages, and more interference in management decisions. On the other hand, mature unions might be preferable to a work force that is in the early stages of unionization, foreshadowing strikes and violence. Unfortunately, the only data on unionization are reported by states.

Employment of low-skilled workers is impeded not only by a wage barrier but also by low education in most mature industrial areas. Metals, machinery, and autos have for decades paid wages for assembly line and other routine work that not only attracted the less educated but also repelled the well educated because the work was so routine and dull. Our statistics do not reveal a local failure to support education, but they do suggest a net migration flow—a brain drain—that beggars the region.

Given the impediments to attracting low-skill, low-wage work, mature industrial areas may have to rely heavily on an import substitution strategy that exploits local markets to the fullest, offsetting any disadvantage in local production costs with transportation cost savings. Manufactured housing produced for sale in the local market and exported only 200 miles or so would meet these conditions, as would most service industries.

A readily expandable local labor supply depends significantly on the work attitude and experience of females. A high female labor force participation rate indicates greater interest in working, and the occupations of actual employment give information on the experience of that labor force. On the other hand, a low female labor force participation rate suggests a greater potential for easy expansion of local employment, particularly within the lower-wage industries and occupations. The effect is reinforced when the unskilled, unemployed females are immobile because they are married to locally employed males. Historically, apparel factories were attracted to steel towns by the ready availability of steel workers' unemployed wives.

Better information than is now available should be used to choose between the options of wage cuts, investments in general education or vocational training, outmigration assistance for those whose skills are no longer employable within the area, and import substitution strategies. But we feel increasing confidence in our research strategy to guide the choice.

POSITION AND PROSPECTS IN PRECISION OPERATIONS

We have tried here to steer a middle course between expecting too little of the local labor force and expecting too much. We are not concerned here with the creation of new high-tech products or processes, only the intelligent and responsible use of expensive high-tech equipment and the faithful following of complex procedures.

Surprisingly, the ratio of precision production, craft, and repair workers to operators, fabricators, and laborers in mature industrial areas is only 0.6 to 0.7—two higher-skilled workers for every three lower-skilled workers—below the national average. Even more critical in the long run, almost one-third of that local labor force lacks a high school diploma, which is also below the national average, and only one-sixth have graduated from college, which is well below. Half the local labor force in the national industrial centers is not well prepared for precision work, much less high-tech research and development.

In sharp contrast to a science city where both scientists and engineers are counted, with extra weight given to those with doctorates, the precision index counts only the local supply of engineers. And to suppress the dominance of college towns, extra weight could be given to the 1974 NSF survey that identified those scientists and engineers employed in business. More current data is reported in the 1980 Census count of engineers.

A college degree may soon be required for most precision work, whether on the factory floor or in the laboratory. In any case, even a first-class vocational education presumes a good high school education or its equivalent.

A second measure of skill is the percentage of the local labor force employed as technologists and technicians (excluding health), which is designed to capture the shift from blue-collar work into "white coat" positions, including CAD-CAM and the constant inspection adjustment and redeployment of robots.

Those making plant or office locational decisions would be strongly attracted to a local labor market that had an excess supply of key skills, that is, one that offered both ample supplies, soft prices, and greater choice of personnel. With careful work, one could approximate this condition with the ratio of current local employment of, say, machinists to number of machinists in the local labor force. An alternative index is the ratio of current local employment of a given skill to the peak employment of that skill within, say, the past two years.

The less mobile the workers, the more important is the disparity in supply between the locality and the nation. Professionals can be recruited from afar; supporting technologists and technicians should be on the spot or nearby. Local commuter colleges such as Cleveland State University, with 99% of its enrollment local residents may not be cosmopolitan but they do display the virtue of graduating students that are likely to remain in the region and therefore return local social investment to the communities.

The rapid growth in the 1950s has left many Midwest industrial areas with a near-retirement-age bulge in age distribution of skilled labor. The slowing growth of the Detroit-area economy is reflected in the percentage of tool and die makers aged 55 and older, an increase from 18.5% in 1970 to 23.7% in 1980. The durable goods boom of the early postwar period led to large hirings of 26- to 40-year-old artisans who are now, 30 years later, near retirement. For comparison, note that only 17% of all workers in that area are 55 years old or over.

Mature industrial areas are in danger of losing their competitive edge in skilled labor. The good news is that unusually heavy retirements will create a large local replacement demand for workers whose skills are still in demand, offsetting in part the lack of local growth.

It may be argued that a high proportion of homeowners in the local metropolitan area is either an asset or a liability in local economic development, depending on whether the homeowners are skilled or unskilled workers. Given that workers who are also home-owners are more inclined to defer moving out of a contracting labor market when they have lost their job, this action on the part of unskilled workers delays the needed contraction in the local labor supply, inflates welfare roles, elevates tax rates, and inhibits economic recovery. But the sluggish outmigration of the skilled workers could become an economic development asset. Professionals, who are much more mobile, whose recruiting network is more efficient, and whose employers often pay their moving expenses, are much less likely to stay on and attract work. Similarly, skilled workers are more likely to buy homes than are either professionals or the unskilled because unlike the former, they do not move about in a national labor market and, unlike the latter, their incomes are adequate to buy homes.

Looking to the next generation, state and local expenditures on education per capita and per dollar of personal income are fair measures of the citizen-taxpayer support of education and a rough measure of the quality of the local schools—perhaps even of the quality of the next generation of labor. And right away, good schools attract parents of school-age children, especially the most educated.

POSITION AND PROSPECTS IN RESEARCH AND DEVELOPMENT

Proficiency in science and technology is seen today as the royal road to local economic recovery, and one of the few open roads to reindustrialization for midwestern urban areas, bypassing their very high wage rates and aging plants. The promotion of science-based industry is, in fact, often seen as almost the only escape from the devastating competition of low-wage, foreign labor.

The most detailed readily available sources of data on the geographic distribution of scientific and technological capabilities are the 1974 and 1975 surveys of the National Science Foundation and the 1980 U.S. Census of Population. The number of doctoral scientists residing and employed in an area, regardless of occupation or employer, is a good general measure of professional excellence in background and training—the availability of the highest level of expertise, whatever the local use of that expertise.

A second attribute selected expands the technical manpower base from doctorates only to all scientists and engineers, but narrows the focus to include only those engaged in research and development work (without distinguishing at this time between basic research, applied research, and development, which the data do permit). A third manpower indicator retains the broader base of all scientists and engineers but shifts the focus to cover only those employed in business and industry. By using all three of these variations, we implicitly give triple weight to a doctoral scientist or engineer engaged in research or development work for a private business employer (for example, a manufacturing firm). And double weight is given to any combination of two of those three characteristics:

(1) doctorates engaged in research at a university,
(2) doctorates employed in business management (including administering research and development), or
(3) nondoctorates engaged in industrial research and development.

This blend of indicators gives only single weight to doctoral professors engaged primarily in teaching and avoids overweighting university faculty and unduly elevating college towns. Educational institutions employ over half of all doctoral scientists compared to only about one-tenth of nondoctorates. We have also guarded against generating very high, misleading ratios produced by dividing modest numbers of scientists or research expenditures by very small populations (such as those of college towns) by confining our early runs to metropolitan areas of one-half million population or more.

Professors do, however, enter the composite index with single weight to reflect their contributions—outside of their classrooms—to the local level of technology, expressed through consulting, among other things. Professorships are generally preferred to research positions in universities; therefore, to have excluded teachers would have eliminated many of the most prestigious and productive scientists and engineers.

The academic base for science and technology is measured here by the number of earned doctorates granted by local universities between 1975 and 1980 in the life sciences, physical sciences, and engineering expressed per 100,000 population. All doctorates in the natural sciences and engineering were weighted equally, although the life sciences produce more doctorates than the physical sciences and 40% of the physical science doctorates are chemists. For special focus, the earned doctorates can be indexed for four engineering fields, six physical sciences, and six life sciences. University research and development can also be expressed by total funds committed to research and development, using NSF survey data covering the 200 largest universities (about 97% of all research and development).

These individual measures of the local scientific and technological base were selected with an eye to their comparability across different scientific disciplines, heterogeneous occupations, and diverse employers. Persons employed and dollars spent are intuitively more deserving of equal weight between industries and places than other measures that come to mind. Articles published and patents granted will probably prove to be very valuable measures at a finer grain, when looking at the local performance in scientific and developmental work within specific disciplines. But when aggregating, if published articles are shorter or more numerous in some fields than others, then those industries and places based on scientific disciplines in which publications are more numerous or on technologies in which patent variations abound will appear to be more productive in science or technology than they really are, and vice versa. Counting persons and dollars seems a safe way to begin. (For greater detail on this path, see REI Review, 1983).

POSITION AND PROSPECTS IN CENTRAL ADMINISTRATION

One of the marks of advanced local economic development is an important role in central administration. Large corporate headquarters become centers of decision making, especially in resource allocation, and gather together the requisite talent and expertise. Multinationals especially tend to transform the local economy into a world crossroads for scientific, technological, and commercial information of all kinds, and to create a climate attractive to able and educated persons. The assembling of specialists in foreign trade, language, customs, and culture broadens and deepens ties with other countries, further linking the local economy to the growth of foreign trade.

A ready source of data on manufacturing headquarters is the *Census of Manufactures*, which reports employment in "central administrative offices and auxiliaries" by both industry and metropolitan area. The data are for the census year 1977; these data have been supplemented by tables from *Fortune* magazine. We have chosen the total *worldwide* employment of each of the 500 largest industrial corporations as a proxy for their economic size and power. By summing employment everywhere for the corporations headquartered in a given metropolitan area, we have created a number that measures roughly the resource allocation decisions made in a given metropolitan area. The *Fortune* Second 500 profile of intermediate-size corporations identifies what may well include the next generation of multinationals—good substitute performers waiting in the wings to rescue the locality when the local stars become injured or abducted in competition or corporate mergers.

The competitive position of a metropolitan area as a center of corporate decision making is enhanced by an ability to draw on the full range of human resources in that local labor market. One measure of the breadth and depth of use of local talent is the proportion of minorities employed as administers, executives, and managers. The percentage of employed females and blacks who were reported in the 1980 Census of Population as managers are proxies for equality of opportunity and unconstrained use of human capital. And, in fact, the percentage of females employed in administrative, executive, and managerial occupations does vary markedly from place to place.

Not only may female managers' contributions be additive and distinctive, but places that offer women greater occupational opportunity tend to have a better chance of attracting and holding men as well as women, especially husband-and-wife teams in which both are highly educated and pursuing specialized professions, that is, key labor in joint supply. The price of professional labor will tend to be lower in places that offer occupational extension ladders to talented females.

The low proportion of female executives and managers, characteristic of Midwest industrial areas (6% compared to a big city average of 8%), may be explained by local specializations in durable manufactures—heavy industry that is less than hospitable to females—but the legacy remains, and this area has a shallower pool of experienced female executives and managers on which to draw. This partly self-imposed development constraint could be especially binding in activities for which females have a comparative advantage, such as in product design and marketing, to capture subtle dimensions on new lifestyles.

In addition to the nature of the work of metal making and metal bending, further study will probably show that heavy-industry towns have drawn many ethnic groups that have looked with less than favor on women working outside the home. And the high wages paid male workers have made it less necessary to add a second income to the household. Perhaps the new generation will eliminate this legacy.

POSITION AND PROSPECTS IN ENTREPRENEURSHIP

It is easy to think a mature economy can escape from dealing with the difficult problems associated with being a high-cost location—high wages and fringe benefits, high taxes, and stringent regulation—by taking refuge in the fact that most employment growth comes from the birth of new firms or the expansion of existing local businesses. If we can grow our own new businesses, why worry about comparative cost disadvantages?

This argument lightly assumes that the local ground is fertile for cultivating new businesses. Constructing a meaningful index of local entrepreneurship will be a challenge. We begin that work by looking at the proportion of local self-employment. With only 4.0% self-employed in 1980, mature centers of heavy industry, such as Cleveland, Chicago, Detroit, and Milwaukee, cluster at the bottom of the list of the 32 largest metropolitan areas. Admittedly, this index includes routine proprietorship—"mom and pop" stores. A complementary index can be added by deriving the proportion of manufacturing activity (employment or number of firms) accounted for by small firms (fewer than 20 employees).

Centers of heavy industry show the legacy of decades of dependence on large corporations. Why should able and energetic young adults—potential entrepreneurs—take risks and work long hours starting up a new business when a degree in engineering or business administration leads to a high salary and a secure job at Standard Oil or General Motors? The critical question becomes whether recreating the local culture of entrepreneurship is easier to accomplish than cutting back those comparatively high costs of production.

Our not very surprising early finding is that the industrial Midwest is below average on almost all measures of entrepreneurship. This would suggest that local economies swing through long cycles of, first, sharp bursts of invention and innovation, typically evolving into periods of aggressive management that build large business organizations—convoluted bureaucracies almost destined to degenerate into the caretaker management—that erode the local industrial base, and finally induce, again, the necessity that mothers invention. Whole communities, mimicking the proverbial family, evolve from rags to riches to rags in three generations. Are there private and public strategies that might accelerate the return of a local climate of entrepreneurship?

The venturesomeness of the local economy is also determined by the variation within the population from which entrepreneurs are drawn. Again, centers of heavy industry report the lowest proportion of females who are self-employed, with, for example, Cleveland and Chicago attaining only about half the average figure of 4.8%.

An important concern is that these measures of local entrepreneurship may be biased against large organizations and thereby underestimate the vigor of the places that host them. It is important to construct a measure of "entrepreneurship," too (creativity and risk-taking within large organizations). For example, the revolutionary Macintosh computer was invented within the billion dollar Apple computer corporation by a very small team handpicked by its founder, Steve Jobs. This think tank was organized so loosely that skeptics dubbed the project "Steve Job's back-to-the-garage-fantasy," an allusion to an attempt to relive the founding of Apple. But it worked. It also worked at a time when Apple's chief competitor created the IBM Personal Computer through virtually the same process.

We may find that fast-growing divisions of large corporations rival the spectacular rates of the most successful small firms, about which there has been so much favorable comment over the past half-dozen years. In fact, small businesses may only seem to account for a disproportionate share of the economic growth. Small firm deaths may be undercounted and internal corporate divisions that are fast growing are averaged with those that are declining.

Deep pools of venture capital—both equity and debt capital—accessible to new small business, are well recognized as a critical element in the growth and development of local economies. Further, venture capital is probably more important to large places than to small ones because large places are more dependent on innovation whereas small places are sustained more by the attraction of branch plants. To the degree that large places have higher costs of production for routine operations, they must rely more on their special ability to perform new tasks that begin high up on the learning curves, and they must plan to replace this familiar work as it becomes simpler with time and experience and filters down through the national system of cities to smaller places where it can be performed by lower-skilled, lower-wage labor.

Seed money for small firms is most likely to be coaxed away from relatives, close friends, and business associates. Therefore, unlike most capital that is raised in national capital markets, venture capital markets are localized, and of special interest and demand special attention in large, mature local economies seeking reindustrialization. San Francisco, Pittsburgh, Chicago, and Boston have, in our measurements, led the way in financing new small businesses, currently running at two to five times the national average.

One can speculate that venture capitalists serve the cause of invention by speeding the transfer of ownership and management of new ventures, forcing the sale of new companies at the earliest point of profitability. The logic here is that the sooner inventors get back to inventing and out of managing, the better for society and the local economy. One might further speculate that good ideas are the truly scarce resource, and aggressive investing in and lending to new businesses may boost failure rates more than local economic development.

Finally, local employment in marketing is used here as an index of local entrepreneurial activity and strength. Marketing is generally one of the weaker skills of small businesses (along with skills in raising capital and in contrast to the typical strength in production).

Marketing contacts and skills are often especially weak in small businesses that have been long-time captive suppliers of large corporations, for example, automotive tool shops and parts suppliers. They have been sustained for decades on passively waiting for their dominant customers to bring in the blueprints for competitive or negotiated price setting.

TOWARD POLICY-BASED RESEARCH

The metropolitan area—defined as the job commuting radius that bounds the local labor market—is such a close approximation of the local economy, and economists feel so comfortable working within this familiar space that they are understandably reluctant to leave this intellectual womb. And these ties are tightened by the superior quantity and quality of data available at the metropolitan-area level over any other state geographic subdivision.

But leave they must if they expect to bridge the gap to policy relevance. We do not need to labor the glaring lack of effective metropolitan area government in the United States. Even the exceptions to the rule—city-county consolidation—usually lag behind events; most of the new suburban growth is in contiguous counties beyond the reach of the new "metropolitan area government" by the time it takes effect. When the metropolitan area analysis is finished, there is simply no one to hand it to—no one who can run with it.

The regional-urban analyst who seeks to contribute to policy-making could carry the work to the state level by extending the analysis to much or most of the urban systems within the state in order to capture the attention of state government. The state also bears the responsibility for the part it plays in arranging the economic furniture within its own house. The distribution of funds for state universities—especially graduate programs in science and engineering—has become the critical variable in the geographic distribution of employment in research and development. Again, state transportation policy, planning, and public works create hubs and networks that sometimes create new centers and always affect the position of existing ones, which act, in turn, to assign headquarters roles to some cities rather than others. Those who allocate state funds for higher education and transportation have more control over the settlement pattern of the state for the decades to come than they know or perhaps welcome.

We have completed the work of composing charts similar to the one included here for the Cleveland area for all the other Ohio metropolitan areas, and for not only 1980 data but for the 1970 data and the decennial change, that is, trends. We have completed an economic-topographic map of Ohio that at least roughly sketches the occupational-functional peaks and valleys of the state as a guide to the official pathfinders in state economic development policymaking and programming. We have now come full circle: beginning by rejecting the state as a true local economy and ending by accepting the need to look to the state as the critical political-economy in local economic development.

INFRASTRUCTURE PLANNING: PAVING THE PROPER PATH

Over the past 10 years, urban observers have become increasingly disturbed by the almost irresistable temptation to divert scarce tax revenues away from the maintenance of public works and toward more visible and politically sensitive public services. Let the bolts in the bridge rust a little more but beware of letting garbage steep a few more days—if you can't see it or smell it, forget it. But pressure is building to force capital accounts on politicians and public managers, to make the aging and deteriorations of public capital more obvious, and perhaps even force the accumulation of depreciation funds for regular replacement.

But it is not merely enough to repair regularly or replace promptly aging capital; old may also mean obsolete. Successful businesses replace obsolete machines with ones of improved design; successful cities should sometimes repair, sometimes relocate, and sometimes transform aging capital. A forward-looking public capital program would capture funds for renewal and invest them in forms and places that will support the economic development path chosen. Urban public investment and land-use planners should be prime users of some form of our five-path approach.

For example, heavy investments in high-speed, infrequent-stop rail transit tend to reinforce their corridors but are unlikely to make major contributions to the journey to work in either routine or precision manufacturing plants. And rail transit may in fact never again be as central to routine clerical operations as it has been historically, assuming that the time has passed when masses of clerical workers must be assembled in dense clusters. Again, even if transit continues to play a key role in getting to offices, it may be more to accommodate managers and professionals who would rather read than drive, and the appropriate new, luxury design may emphasize comfort more than cost.

Finally, this strategic development planning deserves to be carried to a rough judgment on whether local aspirations and expectations in headquarters work applies to industrial or nonindustrial companies, or both. The headquarters of manufacturing companies do not seek downtown sites to nearly the same degree as do those serving banks, other financial activities, or utilities. Manufacturing companies do not interact with each other—and thereby generate the making of trips—to anywhere near the degree that financial institutions do, especially investment banking.

The close ties between suburban manufacturing headquarters and their branches and customers around the country and world suggest the importance of easy and rapid access to airports. But these low-density trip origins typically make a better case for expressways than for rail transit connections, for example in Detroit. Compare, again, a downtown focus with a suburban research university. Clearly, the infrastruture strategy for airport access calls for a clear and careful identification of the local position, prospects, and potential along both the headquarters functional row and its interactions with various manufacturing and nonmanufacturing industry columns.

CROSS-CLASSIFYING INDUSTRIES AND OCCUPATIONS OPENS UP NEW POSSIBILITIES IN INDUSTRY TARGETING

A serious shortcoming of industry targeting as typically practiced is that the development strategist too often acts as if it is necessary to target the whole industry, from the headquarters through the laboratories to the assembly line. The local economy may not have the hub location from which to carry on headquarters work, or enjoying that favored but high-cost location, it may have to accept the decentralization of routine operations in that target industry to smaller, low-wage, low-rent places, or the local university may attract only the research and development to its campus setting. The occupational approach detailed above reminds the developmental strategist that an industry can be disassembled and only part of it targeted to take advantage of the special locational needs and preferences of the many different operations.

Mature industrial areas could, for example, seriously consider some fast-growing industry despite that industry's characteristically low wages for routine operations (for example, surgical and medical instruments with half the auto plant wage) but only if local strategists target the headquarters, research, or pilot plant operations where low-skill wage rates count for little in the location decision. Local development strategists would do well to look to the intersection of a growth industry column with their best shot occupational role. Remember the worldly-wise Mr. Robinson in the movie *The Graduate*, who whispered "plastics" into the ear of Dustin Hoffman? This industry tip still left the ambitious youth to ponder the occupational choice of technologist, manager, or entrepreneur—a path through the laboratory, the accounting class, or cheap loft space.

NEXT STEPS ALONG THE PATHS

We are aware that even five distinct paths to local economic development are too few and too aggregated. We have already begun to separate precision work into the older blue-collar and the newer "white-coat" skills. Science cities can be distinguished from the centers of applied engineering or "advanced manufacturing." Entrepreneurship can be disaggregated into creativity and financial risk-taking, and so forth.

Again, we have barely pioneered the complex way in which local economies use functional strengths or occupations to move between industries, how they "reindustrialize," and the process through which industries develop new occupations and new skills. For example, corporate headquarters strength played through corporate acquisitions may be one of the more traveled paths to new local industries, technologies, and skills.

APPENDIX: PRELIMINARY PROFILE OF THE CLEVELAND SCA

The Cleveland-Akron-Lorain-Elyria consolidated area (CALE) is charted along the five paths to economic development in Figure 14.1. The peer group is the other metropolitan areas of over 1 million population in 1980, and the variables are logically related to the paths. As there are 32 metropolitan areas in the population, CALE has a rank for each of the variables ranging from first to thirty-second, and this rank is where the variables are plotted. Where arrows cluster is an indication of comparative economic strength for that path. CALE's profile is low, high, low, high, low for routine operations, precision operations, research and development, central administration, and entrepreneurship, respectively.

The high relative wages for low-skilled work, both factory and office clerical, combine with the second worst record in work stoppages to make this metropolitan area a high-cost place within which to carry on routine operations of almost any kind. The latter statistic does not indicate which side of the collective bargaining table bears the greater responsibility for the work stoppages.

CALE performs above average in precision operations, and the higher the level of skill required, the higher the local rank. The ratio of blue-collar skilled to semiskilled and unskilled workers is twenty-fourth to twenty-fifth, the percentage of technologists and technicians (the "white coats") is twentieth to twenty-first rank, the percentage of engineers self-reported in the 1980 Census attains to midpoint, and per capita scientists and engineers employed in business attains eleventh rank. The area stands out in the stability of its skilled labor force, with a high proportion of homeowners and a very low quit rate. Not many walk away from a high-wage job in a loose labor market. In short, CALE is above average for precision work given improvements in the high school graduation rate: almost one-third of the local labor force lacks a high school education, placing the area two ranks below average.

In research and development, the absence of the largest state university in this, Ohio's largest metropolitan area, depresses the per capita output of doctorates in science and engineering and the associated research and development expenditures at local colleges and universities. The region still manages to achieve an almost average standing in practicing scientists and engineers and an above average proportion of doctoral scientists and engineers. Even so, the area faces an uphill climb to become a national science city, given its low rank in college graduates: thirtieth of 32.

CALE displays strength in central administration, ranking sixth in the absolute number of headquarters employees reported in the *1977 Census of Manufactures*, and tenth in the percentage of total manufacturing employment in central administrative offices. The area rises to fifth in the absolute number of *Fortune* 500 employees whose work is directed from local headquarters, and fourth (after Pittsburgh, Detroit, and New York) in a relative version of this index: the ratio of the number of employees directed from the area to the population of that metropolitan area. In the *Fortune* Second 500 Industrials, CALE rises to an impressive fourth and second place in absolute and relative rank, respectively. The high performance of CALE in the *Fortune* Industrial 1000 is not matched in the *Fortune* Service 500.

In headquarters support, CALE ranks sixth to seventh in the proportion of managers engaged in marketing, advertising, and public relations, and above average in the percentage working in advertising. In the nonprofit private sector, CALE ranks seventh in the foundation expenditures per capita and ninth in the foundation assets per capita. The principal shortfall is in the low penetration of blacks and females into management. A twenty-eighth rank in the percentage of employed females who are managers may be explained as a legacy of the heavy industries like steel and machinery, but the fact remains that the rising role of the information-based and service industries, ones in which marketing to women or marketing skills of women become more important, could leave CALE far behind.

The low standing in entrepreneurship may be more apparent than real, depending on the weight attached to each of two clusters of indicators. CALE lags far behind its competitors in the percentage of self-employed, whether total, female, or black, but this is a crude measure. The region exhibits average performance in large venture capital firms, and two business indexes of fast-growing companies; CALE had 7 of the 100 largest venture capital firms in 1982, attaining fifteenth rank in both absolute and relative measures—an average performance, but one also far behind San Francisco, Pittsburgh, Chicago, and Boston. As a correlate of its strength in manufacturing and corporate headquarters, CALE is above average in advertising. It is below average in the percentage of designers and even lower ranked in percentage of authors. The latter measures are, like self-employed, indicators of the extent to which individual, low-capital ventures can succeed, and are hallmarks of the new information-based economy—warnings to the Cleveland-Akron-Lorain-Eylria area.

REFERENCE

REI Review (1983) November issue. Urban Center, Cleveland State University.

15

Urban Development in an Advanced Economy

ROYCE HANSON

□THE TRANSFORMATION IN THE structure of the national and international economies is remaking urban economies and revising the ways in which urban areas relate to each other (Noyelle and Stanback, 1983). As the Committee on National Urban Policy has pointed out, a new urban system has emerged, characterized by polarization between relatively few large, diversified "command and control centers" and the larger number of remaining "subordinate centers" (Hanson, 1983).

The economies of the command and control centers are shifting from a manufacturing to a service base. These large metropolitan areas contain most of the corporate headquarters of major multinational corporations; the largest international banks; leading research universities; concentrations of law, media, and financial services firms that serve headquarters functions of business and government; the nation's principal foundations and philanthropies; and the nation's most highly developed communications and transportation systems. These centers are increasingly divesting themselves of routine manufacturing and office functions, first to their own suburbs, and later to subordinate cities in the United States and around the world.

Subordinate centers, on the other hand, tend to be highly specialized in production or consumer sectors of the economy, with little local autonomy in economic decision making. Their labor forces tend to be less diverse and flexible and they are far more susceptible to cyclical booms and busts than the command and control centers (Hanson, 1983; Noyelle and Stanback, 1983). They are especially vulnerable to structural shifts in the economy, with those that house growing industries performing well, and those that are home to declining industries subject to a spiral of decline (Bradbury et al., 1982). Subordinate centers that historically have depended on routine, labor-manufacturing are losing jobs to Third World cities or to other locations in the United States where there is a surplus of low-skilled labor.

The increasing polarization between the command and control centers and the subordinate urban centers is one focus for an urban development strategy appropriate to an advanced economy. This focus is concerned with defining and building on the relationship between these two types of centers. It involves recasting the basic notion of all cities as places where production and distribution occur to one that distinguishes between cities whose basic function is transactions in an international economy from those still tied to a regional resource base or to the performance of production and consumer activities controlled elsewhere. For both kinds of cities, a reexamination of the relationship between urban form and function seems appropriate.

A second focus for urban development strategy should be on the relationship between the sectoral and space economies within an urban area. One of the most prominent features of structural change, even in command and control centers and subordinate centers with growing economies, is the simultaneous appearance of areas of growth and decline. Service sector jobs do not necessarily replace lost manufacturing jobs. Rather, they substitute for them. Laid-off blue-collar workers do not readily enter the white-collar market. If they find new jobs at all, they tend to find jobs that pay less. For these workers especially, and for many others with inadequate education and training to enter a knowledge-based economy, the current growth patterns of the economy do not mean more opportunity, but less.

Within the service sectors themselves, the job market tends to be increasingly split between "smart" jobs and good careers and "dumb" jobs and poor careers (Schwartz and Niekirk, 1983). Access to the smart jobs is restricted to the educated and trained. The dumb jobs increasingly are filled by minority and other low-income women. They also tend to include more part-time work, have less generous benefit packages (if any), and few of the protections for workers that characterized industrial employment for the past 30 years.

In this dual context of changing relationships among urban centers and structural shifts within them, averages are of little help in understanding what is going on. A city with an overall unemployment rate of 4% may experience rates of black male unemployment near 25% and a youth unemployment rate of over 40%. A net increase in jobs may mask a precipitous decline in blue-collar jobs or the fact that the income generated by new jobs may be less than that generated by the lost jobs. Enormous growth in some parts of the area due to new office construction, gentrification, and high-income housing is frequently accompanied by precipitous decline in older industrial districts, and working-class and poor neighborhoods. By and large the command and control centers are experiencing a net gain in this process, but many subordinate centers face net losses.

THE NEW TRANSFORMATION

All of this is to say simply that cities are in the grip of a massive and pervasive economic and social transformation as important to their futures as the industrial revolution was. From a strategic perspective the difference in the two eras is that we can intelligently forsee this revolution and some of its implications for the way in which cities will need to work, whereas we had only the dimmest awareness of what the industrial revolution might bring as it slowly proceeded to influence the function, shape, and size of urban areas.

To deal with this transformation, urban development strategy needs to deal with two realms simultaneously; the mainstream of change and those left behind (Committee on National Urban Policy, 1982; Hanson, 1983). In other words, the most workable strategy is likely to be one that seeks to accelerate the transition of urban economies from their former industrial bases to a more advanced character. At the same time, there is a need to stabilize communities and people so they are able to adapt without undue trauma to the shocks of transition. Stability and acceleration are both essential to effective development strategy. Without stability, those being left behind will mobilize their political resources to slow the process of change. In addition, the toll in human worth and the waste in productive capacity is too immense for a great society to do otherwise. Without acceleration, stability will degenerate into static or declining economies. Thus, a development strategy for urban areas moving into an advanced economy calls for a careful balance between the two realms.

In all economically and socially advanced nations urban development is the product of a guided market, an amlagam of private market choices and public policies. Because of its dependence on infrastructure, urban development is highly sensitive to policy decisions. Civilization and market values follow the development of sewers.

Policy, however, only influences, it does not exercise ultimate control over development. Basically, policy can exclude or encourage; it rarely "causes" development. It can exercise enormous locational leverage, however, and it can direct capital to some sectors of the economy in preference to others through the structuring of investment incentives and penalties.

It is also clear from experience that many policies, particularly at the state and national levels, have a considerable impact on urban development although this impact was neither foreseen not particularly desired (Glickman, 1980; Vaughn, 1977). Thus, broad national economic policies, such as tax policy (tax-free municipal bonds for industrial development, accelerated depreciation, tax credits for historic preservation, deductions for home ownership) can guide capital into specific sectors of the economy and into particular locations within urban areas. Monetary and credit policy directly affects the rate of urban development. Budget, trade, and regulatory policies influence both investment and disinvestment. Direct and indirect subsidies to particular industries, such as agriculture, transportation, defense, and scientific and technological research and development are not space-neutral. They are of enormous benefit to some types of urban areas and of disadvantage to others. At the state level, education policies, infrastructure decisions, and revenue-sharing formulas are all important factors in the definition of the "local" market for urban development.

At the local level, planning, zoning and subdivision policies, infrastructure decisions, the use of eminent domain, and direct subsidies all color the "market." Although in theory price is the arbiter of the market, in reality price reflects the fusion of public and private interests in any particular development opportunity (Hanson, 1974). The problem of urban development strategy for an advanced economy is one of both economics and politics.

A NEW DEVELOPMENT STRATEGY

In this light, what must an urban development strategy do? It must, of course, be concerned with the location of capital investments, both among the components of the urban system and within specific metropolitan areas. This suggests roles for state and national governments as well as local government.

There is a national interest in investments that contribute to overall national growth. In one sense, the national government could care less where those investments are made so long as they contribute equally to the growth of the gross national product. But not all investments are equal from a national viewpoint. Investments in growing sectors of the economy are likely to have a greater stimulative effect on the whole system than those made in declining or static sectors.

It makes good sense to know as much as possible about the effect of general economic policies on sectors of the economy, and incidentally, their consequences for different kinds of urban areas. Deregulation of oil prices, for example, had a substantial effect on the economies of several Sun Belt cities, just as the growth in defense expenditures has had on others. Deregulation of airline travel has generally benefited command and control centers that have high-volume international airports, and has caused some economic harm to subordinate centers that are harder to serve without the cross-subsidy from the bigger markets. Although bank deregulation has only begun to have an impact on specific urbans areas, there is strong reason to expect that it will result in increased concentration of the banking system in command and control centers, and that this will in turn have an effect on investments made in many subordinate centers.

All of these policies affect the relationships among parts of the urban system.The question for the national government is whether that is the intention of the policy or whether it is an inadvertant or unanticipated result of policies that were made innocent of any spatial perspective. If the later is the case, then it probably makes sense to consider spatial consequences that can be anticipated before policies are adopted rather than later introduce countervailing policies to correct the imperfections of the first.

There is an even more important reason for national concern with relationships among the centers in the urban system. It is now an international system, with American cities in direct competition not just with each other, but with locations throughout the world for different economic functions. Because an advanced economy is more knowledge based than an industrial economy, it matters greatly where the nodes of knowledge are located. The nation that is best able to develop such centers is most likely to have the capacity to keep its economy at the cutting edge of innovation and thereby retain a continuing competitive advantage in the international economy. Thus, there is a national interest in developing a network of world-class cities containing agglomerations of command and control functions. It would also appear to be sound policy to think of the rest of the urban system as both supporting the command and control centers and as containing a number of places that can be developed over time into additional regional, national, and international centers housing command and control functions.

A second level of national interest in urban development should arise from the problem that areas that lag well behind the average national growth rate pose for the overall economy. Such areas represent lost productivity and income to the nation. Although the argument is frequently made that people will vote with their feet and move from chronically depressed areas (President's Commission, 1980; U.S. Department of Housing and Urban Development, 1982), there is strong countervailing evidence that at least in the short to mid-term, it is easier to move capital in some sectors than labor (Bluestone and Harrison, 1982; Clark, 1983). Also, from a perspective of efficiency of resources, it makes some sense to use existing urban infrastructure where possible rather than require new public investments in other locations.

At a minimum, federal policy should not, as it now occasionally does, actually provide incentives and inducements for disinvestment in some urban areas and the transfer of capital to other places where public subsidies are then used to reestablish and modernize facilities.

Although the federal government may be ambivalent about its role in the spatial distribution of capital, the states have "got religion" on the subject. The importance of capital flows to states is illustrated by the intense "courting" of the new General Motors Saturn automobile plant by the governors of almost a dozen states. Each governor put together a package of subsidies—land, interest, tax credits or abatement, and so on—to attract the 6000 jobs the plant would provide and the considerable multiplier effect it would have on the local economy in which it located. Several of the governors even appeared on the *Phil Donahue Show* to publicly compete in the bidding war that ultimately would have little effect on the location decision of corporate management, inasmuch as any package of state subsidies would produce such a small marginal benefit that the result would be a windfall to the company from the "winning" state, which would probably have been chosen even if it offered nothing.

THE NEW MEANING OF MANUFACTURING

Indiscriminately chasing smokestacks and silicon chips, state policies tend to focus on job creation, but show little real appreciation of what is required to attract job-creating or wealth-generating capital. Manufacturing will in most instances create fewer direct jobs than services. In an advanced economy, manufacturing is increasingly capital intensive. This does not mean that modern manufacturing is a poor investment for states and communities, however. As it becomes more knowledge intensive, manufacturing—like agriculture—will employ fewer people directly, but will demand a more highly educated and trained labor force and will generate demand for more service jobs. But in an advanced economy, "manufacturing" does not necessarily mean making things. It may mean headquarters and research facilities in which people plan, manage, design, market, and support other people in other places who actually make the products.

In devising development strategies, states need to think about the production cycle (Rees, 1980) and what happens as industries go through the processes of development, routinization of production functions, and dispersion of activities. The need to apply the knowledge of their own economies that exists could be readily developed to identify those places where their intervention is most likely to have some real impact on location decisions. In general, it would appear that as the knowledge requirements for both manufacturing and services grow, the quality of the labor force is of paramount importance. As states are constitutionally responsible for their education systems, the development of a labor force prepared for work in an advanced economy would be the wisest investment in attracting growing sectors. Such a labor force will need to be not only able to enter the new economy, but the state education and training system will need to be designed so that the labor force can keep up with rapid technological and institutional change, and be resilient to the need to work in a different kind of career system than has heretofore characterized most industrial labor systems. That means ability to change jobs and careers several times in a work life, and to move laterally from firm to firm or even between sectors. In a knowledge-based economy, cheap labor will not be as important as good labor.

States that are losing industrial jobs, however, have to deal politically with the short run if they are to have the luxury of even thinking about the long-run strategies they should employ. Under intense pressure from communities and workers that are losing their jobs, states tend to react by seeking replacement firms like those that have closed or moved away, or by trying to develop resource-based jobs. Once again, this often is a losing strategy, as it involves propping up obsolete or shrinking sectors with subsidies. An alternative approach is to analyze the generic skills that dislocated workers have and seek firms that can use those skills even though they may be unlike the firms that have left. Structural unemployment, unlike cyclical unemployment, does not end when the general economy snaps back. Thus, some programs for the physical relocation of excess workers may be necessary. This requires both national and state programs for retraining, job information,and relocation subsidies. At the national level, it also requires the nationalization of the welfare system so that workers can move without penalty of losing eligibility for benefits if they are again unemployed after a short time.

Federal and state urban development subsidies are now available in two principal forms. The largest subsidies are indirect, in the form of tax expenditures such as tax-exempt state and local industrial development bonds, or various kinds of tax abatements and credits. In most states, industrial revenue bonds are indiscriminately available. Local governments issue the bonds, now almost routinely, as an inducement for development regardless of whether the development fits any strategic plan for the local economy. In some cases they are as easy for a developer to obtain as part of a financial package as are connections to municipal water. (They may be easier, as the developer may not have access to public water systems.)

Direct subsidies are smaller in amount, but are often critical elements in specific development schemes. The most significant of the direct subsidies is the Urban Development Action Grant (UDAG) available on a "retail" basis from the federal government. A city must demonstrate that it has a viable project that cannot be developed "but for" the federal grant, which may be used in various ways to leverage the private investment. The grant application must meet a number of federal criteria relating to private leverage, jobs created, and type of community served, but the decision is essentially made on the basis of these criteria and no small amount of political pressure from the city through its congressional delegation and the White House, by the U.S. Department of Housing and Urban Development. The federal government also provides other indirect aid to development through the Community Development Block Grant and through general revenue sharing, but the actual disposition of these grants to projects is determined at the local level. A small amount of development assistance also remains available through the Economic Development Administration. Job training funds are now dispenses through the states under the Job Training Partnership Act.

It is at the local level that the demand for subsidies is most fierce. Within metropolitan areas, central cities are locked in combat with their suburbs for revenue and job-producing development. Cities within the same state are frequently competitors for the same project, and competition between cities in different states remains intense.

In addition to serving as conduits for federal and state subsidies such as UDAG, many cities have authority within state or federal guidelines to issue revenue bonds for economic development. A number of states have recently authorized or expanded other subsidy programs, such as tax increment financing districts, which allow the city to freeze taxes at the "before-development" level in a district and use the increment of increased revenue generated by the development to repay principal and interest on bonds, the proceeds from which may be used to provide public improvements, land, or financing for the development, thereby reducing its cost.

The object of the whole system of subsidies, at least to the extent they were consciously designed to assist the urban development process, is to redress the market imperfections and the adverse effects of other policies that produce competitive disadvantages for some areas in attracting development. The basic problem with the current subsidy system is that it equals less than the sum of its parts.

Aside from UDAG the system is largely untargeted. The political milieu in which it functions responds to the "iron law of dispersion" (Dommel et al., 1978), which suggests that any targeted subsidy program will ultimately be undermined by pressure to make it available to all areas. Moreover, with the exception of a few cities like Baltimore, there is no sense of urban development strategy at any level of government against which the utility of specific policies or packages of them can be assessed.

THE ROLE OF STATE STRATEGIES

A state strategy, for example, might well be concerned with a tailored approach to the different kinds of urban areas within the state. It might restrict the use of IRB financing to those areas where there is a real need to subsidize financial packages to make costs of development competitive with other locations. In command and control centers, it might deliberately use state subsidies and investments in infrastructure and services to help establish or retain certain functions, such as headquarters, banking, and research and development functions in particular districts. Generally, it would seem prudent to insist that it be limited to export or import substitute industries and to firms that hold promise of growing or of encouraging the formation and growth of other firms, preferably in the vicinity.

Working in this way requires a fairly explicit state economic development strategy that takes account of the differences among urban areas and the functions they perform. It would recognize that different urban areas have different mixes of industries and occupations on which to build future investments and linkages among command and control and subordinate centers. It means having a sense of how a state's urban areas relate not only to each other, but to other places in the region, the nation, and the world.

Within urban areas, development policies have been fragmented, with each jurisdiction pursuing its own interests, if possible at the expense of its neighbors. In part, this is a function of dependence on revenues from locally based activities. One way of overcoming this situation is state action to equalize local fiscal capacities by systems of intergovernmental revenue transfers (Advisory Commission on Intergovernmental Relations, 1982).

In an advanced economy, interlocal competition within a metropolitan area makes even less sense than it made in an industrial city setting. An agglomeration of advanced services, headquarters activity, research centers, cultural and recreation facilities, and health services is critical to the vitality of the entire area. Most command and control centers need good access to an international airport from a variety of activity centers containing headquarters and related services. The area needs to be structured to facilitate access by advanced telecommunications systems, including fiber optics and satellites. That linked polynucleated structure can, of course, occur by chance and the force of the market. It can, however, be planned at the metropolitan scale and brought into being far faster through coordinated implementation by all concerned levels of government.

The need for coordinated action is even more urgent for subordinate centers that suffer from lack of economic autonomy to start with, and find economic decision power further fragmented by political boundaries that separate political and economic power. Cities that have housed production functions tend to be disaggregated to start with. Facilities are distributed widely about the metropolitan area, often in different industrial suburbs. Downtown areas are weak. Private leadership is often hard to mobilize, especially to foster change in economic functions. Services, particularly in the central city, are often inadequate. Amenities may be limited.Once some of the key industries decline, the problem is to maintain a critical mass of anything that can be the basis for new economic development.

In some cases these urban areas can be modernized and continue their historic production functions. Even if this can happen, a substantial part of the labor force must be retrained, and in all likelihood a major upgrading of the education system will be needed to develop an effective labor force for the future. In other cases old cities must be transformed so they can perform new functions. This will entail strengthening those sectors that can grow corporate services, such as headquarters activities, that relate not only to local markets but to the broader international market, and strong nonprofit sectors such as health or education. Stamford, Connecticut, is a city that has been transformed from an industrial center to a service center through a strategy of encouraging investments in services as old industries declined. Baltimore is another city that has changed its function in the urban system through using public policy over a long period to foster certain kinds of activities that had potential for shifting the economic mix of the metropolis. Pittsburgh, Akron, and Denver have also been going through metamorphoses, changing their centers of gravity, and moving toward assumption of new roles in the advanced economy that are quite different from those they played in the industrial era. It is increasingly clear that the critical factor in urban development in an advanced economy is not whether a city is located in the Sun Belt or the Frost Belt, but the functions that it has performed in the past and those that it is capable of performing in the future. And the transformation of cities like Dallas, San Antonio, and Atlanta or Charlotte in the Sun Belt are not pure market phenomona unassisted by public policy. Similarly, the revitalization of Indianapolis, Baltimore, Boston, and the emergence of Columbus as a major service center were not aberrations in a cold climate but products of conscious strategies. In some cases, government played a fairly passive role, primarily endorsing the strategic decisions of the private and independent sector leadership.In other cases, government was itself the leading strategist. Interestingly, Frost Belt cities continue to retain more than their national "share" of growth in the economic functions that produce command and control centers. And for the most part, they continue to have a considerable competitive advantage over most Sun Belt cities in a number of important factors essential to growth in an advanced economy; skilled labor forces, well developed infrastructure, good education systems and other public services, and strong cultural institutions and other nonprofit institutions.

THE NEED FOR PUBLIC/PRIVATE COOPERATION

Where transitions have already occurred or seem to be successfully under way, a critical factor has been an effective system of cooperation between the public and private sectors (Fosler and Berger, 1982). This suggests that urban development strategy needs to pay attention to institutional development as well as purely economic opportunities. In fact, economic opportunities can be created where none existed as a result of effective planning in which both sectors participate and share responsibility.

To date, most public-private cooperation has focused on projects. The harder task is to maintain a long-range strategic planning and management effort that can make responsible decisions about helping some industries decline gracefully while others are given encouragement to establish themselves and grow. There is also a substantial challenge to the private sector to see cooperation as more than an opportunity to achieve short-term political objectives, such as lower local or state taxes. A sense of the city as a seat of leadership in an advanced economy suggests an agenda that is hard for either sector to accept and share with a traditional antagonist. It includes the need to reform the education system of most cities, which may involve fundamental redesign of the public school system to increase competition, accountability, and investments, especially in the education and training of the disadvantaged populations of the city. It may also require the redesign of other services to improve their efficiency and equity. These decisions are hard for the public sector, with its established constituencies to accept.

But there is another set of decisions that will be hard for the private sector to accept in many urban areas. Industries, workers, and communities will need to work more closely and cooperatively to avoid the local calamities of economic transformation. This means planning in advance for modernization of plants where possible and for their closing where it is inevitable by training workers for different jobs and occupations. It may also mean that a community is entitled to more, by virtue of its subsidies for a firm's location, than complaints about the business climate. Public investments in industry are also entitled to a return, in a sense, of corporate responsibility to the community and to the workers who invest substantial parts of their lives in the firm. This return may be in equity, some degree of codetermination, or programs that help workers and the community prepare for shifts in technology, products, and processes.

Finally, urban development strategies at the local level must take account of the growing importance of the independent sector. It is not only an important intermediary between the public and private sectors, making it possible to do things that neither could do alone, it is also a major, independent force in the local economy. It provides an alternative means of delivering important community services. It is, through philanthropic trusts, a major source of investment in social programs, services, amenities, and the culture; infrastructure of the city. It is a vehicle for much of the arts and entertainment available to the public. And it is an important part of the institutional system for health, education, and research. The independent sector is in its own right a significant economic sector in the economically advanced city. It is a growing sector, especially in the arts, health, and some aspects of education. It makes a qualitative difference in the role an urban area can perform.

CONCLUSION

Urban development in an advanced economy calls for a different strategy than any level of government has systematically used. The stakes are too high to rely on the chance that some places will develop coherent strategies from the disparate grab bag of programs that are available to meet the opportunities that present themselves. The strategic priorities seem fairly clear.

On the acceleration of change side of the strategy, national policy must ensure that capital is available to modernize industries and to stimulate the development of sectors that promise to be strong, long-term competitors in international markets. National government and state governments might also make a limited amount of leverage capital available on a "wholesale" basis to regions and localities that have developed reasonable strategies, tailored to their circumstances, for facilitating the transition of their economies toward new functions.

The focus of this kind of development stimulation is one giving communities the capacity to seize opportunities rather than to ameliorate distress. It is also important to have available sufficient capital for development and improvement of the infrastructure essential to the efficient functioning of an advanced economy.

The most important aspect of long-term strategy is the development of a competent and resilient labor force in the nation's urban areas. This will require major new investments in elementary, secondary, and continuing education by all levels of government. It will also probably entail some fairly basic restructuring of the educational system in the United States, involving greater participation in education and training by the private and nonprofit sectors to make it possible for currently disadvantaged urban populations to be able to enter the labor force, to progress in it, and to respond to changing demands for labor as the economy continues to be restructured. National policy will also need to be adjusted to facilitate labor mobility through nationalization of benefit and welfare systems and the development of a national labor market information system.

On the stabilization side of urban development strategy, it is important to develop policies and processes that provide time for workers and communities to adjust to structural changes in their local economies, so that workers can be retrained and new kinds of employment developed. To avoid crises in services at the moment of their greatest need, federal and state intergovernmental fiscal transfers should be reconfigured to equalize fiscal capacities of localities. In some communities, especially older urban areas and others facing large-scale unemployment, transitional public employment programs might be used to upgrade local services and facilities as an important aspect of overall development strategy, as a means of training some marginal workers for entry into the economic mainstream, and in a few cases as an employer of last resort.

Finally, urban development strategy for an advanced economy needs to be more conscious than ever of the kind of institutional system it fosters to promote and manage the transition to new functions. An advanced economy clearly is a mixed economy. Markets and policies are inseparably entwined. This makes public-private-independent sector cooperation essential. The form such cooperation takes will depend on local history and circumstances, but its existence will be more than a local concern. State and national development policies should be designed to encourage and nurture experiments that place particular attention in the establishment of alliances that are able to focus on more than projects and look to the formulation and implementation of long-term strategies to improve the quality of the total physical, cultural, and economic environment for development.

The kind of strategy outlined here, the kind called for by the Committee on National Urban Policy, requires a conviction that urban centers are themselves important. They are the places where the advanced economy will in fact work. Thus, they are an integral part of the international competition for those functions that will dominate the new economic system. In this system, firms and other employers do not choose to locate only in nations. They will still have to be somewhere specific, and that somewhere will almost certainly be an existing or newly developed urban area. The country that has the most attractive urban system for serving the new economy will be the country that will take and keep the leading edge of economic growth.

REFERENCES

Advisory Commission on Intergovernmental Relations (1982) Tax Capacity of the Fifty States: Methodology and Estimates. Washington, DC: Author.

BRADBURY, K., A. DOWNS, and K. SMALL (1982) Urban Decline and the Future of American Cities. Washington, DC: The Brookings Institution.

BLUESTONE, B. and B. HARRISON (1982) The Deindustrialization of América: Plant Closings, Community Abandonment and the Dismantling of Basic Industry. New York: Basic Books.

CLARK, G. (1983) Interregional Migration: National Policy and Social Justice. Totowa, NJ: Rowman & Allanheld.

Committee on National Urban Policy (1982) Critical Issues in National Urban Policy. Washington, DC: National Academy Press.

DOMMEL, P., R. NATHAN, S. LIEBSHUTZ, M. WRIGHTSON, and associates (1978) Decentralizing Community Development. Washington, DC: U.S. Department of Housing and Urban Development.

FOSLER, R. and R. BERGER [eds.] (1982) Public-Private Partnerships in American Cities: Seven Case Studies. Lexington, MA: Lexington Books.

GLICKMAN, N. [ed.] (1980) The Urban Impacts of Federal Policies. Baltimore, MD: Johns Hopkins University Press.

HANSON, R. [ed.] (1983) Rethinking National Urban Policy: Urban Development in an Advanced Economy. Washington, DC: National Academy Press.

HANSON, R. (1974) "Land development and metropolitan reform," in L. Wingo (ed.) Reform as Reorganization. Baltimore, MD: Johns Hopkins University Press.

NOYELLE, T. and T. STANBACK (1983) Economic Transformation of American Cities. Totowa, NJ: Allanheld & Rowman.

President's Commission for a National Agenda for the Eighties (1980) Urban America in the Eighties: Perspectives and Prospects. Washington, DC: Government Printing Office.

REES, J. (1980) Government Policy and Industrial Location in the United States. State and Local Financial Adjustments in a Changing Economy. Washington, DC: U.S. Congress, Joint Economic Committee, Government Printing Office.

SCHWARTZ, G. and A. NEIKIRK [eds.] (1984) The Work Revolution. New York: Rawson Associates, Inc.

VAUGHAN, R. (1977) The Urban Impacts of Federal Policies, vol. 2: Economic Development. Santa Monica, CA: Rand Corporation.

U.S. Department of Housing and Urban Development (1982) President's National Urban Policy Report, 1982. Washington, DC: Author.

16

Integrating Urban and Industrial Employment Policies

JOHN P. BLAIR and TRAN HUU DUNG

□THERE ARE THREE approaches to revitalizing urban economies: (1) macro economic policies, (2) industrial policies, and (3) urban policies. Macro economic policies suggest that urban employment problems can best be addressed by overall growth. Hence, the intent of these policies is "relatively neutral among industrial sectors" (Jasinowski, 1980) and geographic areas. Industrial policies attempt to stimulate employment by encouraging growth in particular industries, specifically the manufacturing sector. Urban employment policies consist of programs intended to make urban areas more attractive sites for job creation or to help relocate unemployed workers.[1]

The purpose of this chapter is to compare the three employment approaches and to develop a framework for policy coordination. The first section examines the composition of growth among the northern urban areas using simulation techniques. It shows that no single approach is likely to provide a sufficient level of urban growth. The second section compares urban and industrial approaches and establishes a basis for coordination between them. The final section shows how specific programs fit the policy paradigm developed in the first two sections.

POLICY SIMULATIONS

This section examines the composition of growth in 14 major northern industrial urban counties.[2] Using shift-share analysis, the growth factors for the sample of counties were divided into three components:[3]

(1) *The National Share* shows how employment in a sector would have changed if this sector had grown at the overall national growth rate.
(2) *The Mix Component* shows the employment increase (or decrease) above (or below) its national share due to particular industries growing more (or less) rapidly than the overall national growth rate.
(3) *The Competitive Component* shows the employment increase (or decrease) due to particular local industries growing more (or less) rapidly than the same industry grew nationwide.

These components parallel the three major policy approaches—macro, industrial, and urban policies. Aggregate economic policies—monetary and fiscal policies—are intended to affect directly the national share component. Industrial policies are designed to increase sectoral employment by stimulative (specific) industries' growth. Thus, industrial policies will directly alter the mix component and, indirectly, the aggregate growth rate. The success of urban employment policies, which are intended to make urban locations more attractive, will be reflected in an increased competitive component.

Table 16.1 shows the results of the shift-share analysis for non-governmental employment in the major northern industrial counties. For illustration, consider the FIRE—finance, insurance, and real estate—sector, row 8. Between 1978 and 1982, combined FIRE employment increased by 73,567. Had this sector grown at the overall national growth rate of 5.7%, the increase would have been 52,864. What accounts for the growth above the aggregate shares? Nationwide, FIRE employment grew more rapidly than other industries in the nation. Thus, the mix component indicates that 56,636 jobs above the national share component could be attributed to the nationwide fast growth of FIRE. However,growth is less rapid in the urban counties than in the rest of the country. Had urban growth of FIRE been at the same rate as the national, 35,945 more jobs would have resulted. Thus, the competitive component is negative. The net shift (or variance from the national share) is the sum of the mix and competitive components.

Table 16.1 indicates that the urban counties had negative competitive components for all sectors except communications. For a variety of reasons described elsewhere (Leven, 1978), urban environments were not as conducive to employment growth as the rest of the nation. The mix components indicate that slow national manufacturing growth retarded northern urban growth. But overall, the positive mix component indicates that the economic base of the northern counties

TABLE 16.1
Shift and Share Analysis of Combined Urban Counties, 1978–1982

	Actual Change Regional R82-R78	*Expected Overall National Average "Share"*	*Shift*	
			Mix Compo-nent	*Compe-titive Compo-nent*
(1) Totals	-240,814	489,424	99,689	-829,927
(2) Manufacturing	-473,723	149,900	-282,623	-341,000
Food	-19,285	7,868	-14,647	-12,506
Textile	-5,056	886	-3,375	-2,567
Apparel	-31,630	9,646	-27,480	-13,795
Furniture	10,287	606	-1,727	11,408
Paper products	-3,706	2,866	-6,144	-428
Printing	-1,640	13,773	7,543	-22,956
Chemicals	-4,546	4,326	-5,567	-3,305
Rubber and plastics	-26,117	4,313	-10,751	-19,679
Leather products	-5,573	1,225	-4,413	-2,386
Primary metals	-73,177	11,574	-48,739	-36,012
Fabricated Metals	-62,022	15,125	-35,384	-41,763
Machinery	-44,046	16,070	-3,366	-56,750
Electrical Equipment	-34,295	11,744	4,421	-50,460
Transportation	-72,361	11,158	-33,354	-50,165
Instruments	-9,887	6,853	-674	-16,066
(3) All other manufacturing	-90,669	31,865	-83,581	-38,953
(4) Transportation, and other public utilities	-36,100	34,146	4,775	-75,020
(5) Communication	22,711	7,948	6,569	8,194
(6) Wholesale trade	-33,597	38,483	16,977	-89,057
(7) Retail trade	-79,218	80,952	-2,553	-157,617
(8) Fire	73,567	52,864	56,637	-35,934
(9) Services	285,546	125,132	299,907	-139,493

is not disproportionately concentrated in slow-growth industries. On the contrary, these areas enjoy a slightly disproportionate concentration in fast-growth industries.

Historically, urban area have had positive mix components and negative competitive components. One explanation has been that new industries on the rapid expansion portion of the growth curve benefited from urban agglomeration economies (Blair, 1974). As production became routinized, mature industries would spin off to less urbanized areas where production costs were lower. The almost totally negative competitive components that far outweight the modest mix component in the sample suggest that the northern industrial urban areas continue to spin off activities but have lost some of their attractiveness as economic incubators. Perhaps the urban orientation of fast-growth sectors has shifted to the space, computer, electronic, and recreational cities of the South and West.

The manufacturing sector suffered the most severe decline in employment. Furthermore, workers in the manufacturing sector face a double adjustment problem when they lose their jobs. First, they will probably be forced to change industries and occupations because manufacturing is experiencing slow growth nationwide. (Manufacturing employment actually declined between 1978–1982.) Second, because of the poor competitive component, displaced manufacturing workers in urban area are less likely to find employment locally. Hence, they may also have to relocate. So it is not surprising that manufacturing sectors are a primary concern of industrial policy proponents.

During the period 1978–1982, the labor force grew by about 7.5%, or slightly less than 2% annually. Assuming a roughly comparable labor force growth rate in the urban county sample, the urban counties sample should have generated an estimated 643,979 jobs during the period to provide employment opportunities without forced outmigration. The shift-share framework can be used to stimulate each of the three "pure" policy approaches to attain this goal. Table 16.2 summarizes the findings.

Suppose that each sector maintained its relative growth rate and the urban counties maintained the same competitive position relative to other locations. Now let the national growth rate double from 5.7% to 11.4%. Would the improved macro economic growth have been enough to provide sufficient jobs to accommodate urban labor force growth rate thus avoiding forced outmigration? Table 16.2 indicates that even if macro economic growth had been doubled, urban economies would have had a shortfall of 390,262 jobs, far from enough to provide adequate jobs in the urban areas. The difference between the simulated number of jobs created and the urban labor force increase must be balanced by unemployment or net outmigration.

Turn now to a simulated industrial policy. Suppose an industrial policy were successful in bringing the employment growth of the

TABLE 16.2
Simulated Policy Impact on Urban Counties

	Sufficient Growth	*Simulated Change*	=	*National Share*	+	*Industrial Mix*	+	*Competitive Component*	*Shortfall*
National employment growth rate (5.7%) doubled	643,979	253,717		978,848		104.753		-829,926	390,262
Industry sector employment growth at national rate (competitive component = 0)	643,979	589,111		489,423		99,688		0	54,868
Manufacturing employment growth at national average	643,979	-240,815		489,423		99,688		-829,926	884,794

SOURCE: U.S. Bureau of the Census (1984).
NOTE: The sufficient growth rate of 7.5% is the rate at which jobs would have to be created nationally to keep pace with the increase in the labor force between 1977 and 1982. The rate is computed by summing the percentage changes in the labor force between each year.

manufacturing sectors up to the national average during the 1978–1982 period without slowing employment growth in the other sectors.[4] The northern urban counties would still have lost 240,815 jobs. The difference between the simulated industrial policy and the sufficient growth level is 884,794 jobs.

Finally, consider an urban policy. According to our simulation, even if a policy were successful enough to reduce the urban county competitive component to zero for every sector there would still be a shortfall of 54,868 jobs. However, this case comes closest to generating the 643,979 jobs that would be sufficient given labor force growth.

Summing up, the simulations suggest that no single policy approach in its pure form could generate the "no forced outmigration or excess unemployment" rate of growth. Put differently, if excess urban unemployment is to be avoided, none of the three strategies discussed is sufficient without being supplemented by relocation efforts. Further, note that the simulations do not compare the costs of each approach. Hence, though the pure urban policy approach came closest to achieving the sufficient growth level, a combination of approaches might provide a cheaper alternative.

URBAN AND INDUSTRIAL POLICY

Urban programs and industrial programs address employment problems from complementary perspectives: The former are concerned with the geographic concentrations of unemployment, the latter with unemployment or underemployment due to industrial shifts. Likewise, urban-oriented solutions involve creating location-specific jobs or encouraging individuals to move to places with job opportunities, whereas industrial approaches involve creating industry-specific jobs or training individuals for occupations that offer good job opportunity.

SIMILARITIES BETWEEN URBAN AND INDUSTRIAL POLICIES

Analyses of urban and industrial policy divide into policy/institutional and theoretical. The policy/institutionalist literature generally incorporate many real features of a mixed capitalist system; its recommendations more often than not include specific programs of action. The theoretical literature relies more on abstract economic models in which institutions are kept in the background. However, there is a remarkable similarity in the way urban and industrial policies are viewed in these two bodies of literature. Both sets of policies are considered structuralist in nature, and they both focus on transitional problems that are traceable to wage rigidities and resource immobility.

STRUCTURALIST

Both urban and industrial policy ares structuralist solutions to unemployment. They both assert that macro economic policy by itself is insufficient. Furthermore, both approaches presume that the effectiveness of structural adjustments requires an adequate aggregate demand for labor. The 1982 President's National Urban Policy Report stated unequivocally that national economic health was a prerequisite for a successful urban policy (p. 1). Likewise, supporters of industrial policy on the Joint Economic Committee stated that: "[Industrial policies] are a poor use of resources in a general climate of stagnation" (Joint Economic Committee, 1984: 143).

Under many circumstances, structural changes can improve employment prospects even during periods of insufficient aggregate demand. Even among those who believe in the usefulness of structural policies during periods of high unemployment there is a consensus that targeted incentives are more likely to be inefficient unless the economy is near full employment (perhaps with the help of macro policies).

TRANSITIONALIST

Both urban and industrial policy advocates are concerned about the need to mitigate transitional problems that have possibly been aggravated by the quickened pace of economic change. The 1984 *Urban Policy Report* stated that although some cities (such as New York and Boston) have been transformed to a service-based economy, other cities apparently need urban-oriented programs.

> Cities have had different degrees of success in dealing with their structural problems. The [Reagan] Administration has maintained several economic development programs to aid cities in adapting to these changes [President's National Urban Policy Report, 1984: 72].

Similarly, industrial policy proponents are concerned about the inability of workers to cope with the change from manufacturing to service and high-tech occupations (Blumenthal, 1983; Reich, 1983:

219). They emphasized the need for industrial policy to facilitate transitions "that growth makes inevitable." This is the essence of the Organization for Economic Development and Cooperation (OEDC, 1983: 8-9) support for limited, "positive" adjustment policies.

The transitionalist emphasis reflects the fear that urban and industrial programs may institutionalize inefficiency by propping up obsolete activities. In this respect, industrial policies are considered by some to be different from urban policies because industrial development may contribute to overall revitalization whereas urban policy is, by implication, a drag on overall growth (Schwartz, 1980: 16-17). Urban programs could impede aggregate economic development (James, 1984). Proponents of urban policy, of course, argue that creating opportunities in distressed cities will have beneficial consequences for national growth.

WAGE RIGIDITIES

Both urban and industrial approaches generally recognize that shifts in the demand for labor coupled with sticky, downward wages and prices contributed to unemployment, the standard prognosis of most theoretical models. The view is that employment could always be maintained if wages quickly adjust to any change in demand for the industry's or city's output. Put differently, unemployment is the direct result of the failure of wages to adjust, which in turn can be attributed to various institutional factors such as customs, union contracts, minimum wage laws, and so forth.

Jacobs (1984) speculated that the favorable performance of the Pacific Rim Trading nations and the cities in those nations can be attributed to indirect wage and price flexibility. These smaller nations have an export sector dominated by the output of the major city. If demand slackens, the real price of the product and wages can fall as the value of the domestic currency falls. The situation in smaller, one-city dominated economies differs from that in the United States. If the demand for autos made in Detroit falls, the impact on the value of the dollar will be too small to lower wages and prices, so the decrease demand will result in unemployment.

RESOURCE MOBILITY

In spite of the different reasons for the initial decline in labor demand, the failure of wages to fall sufficiently results in unemployment. As neither the urban nor industrial approach places much faith in wage or price reduction solutions, encouraging resource movements into or out of the declining area or industry emerge as the main remedy. If resources were perfectly mobile, no policy would be needed. In fact, some analysts oppose industrial policy on the grounds that government intervention will prevent "resources from flowing" to more efficient uses (Muller, 1981; Premus and Bradford, 1984: 58).

Inadequate resource (particularly labor) mobility contributes to the need for policy intervention in both approaches. Urban policies are concerned with geographic immobility. Industrial policies are concerned with occupational immobility. Urban-oriented solutions to unemployment encourage individuals to locate in places of opportunity and attract job-creating capital or (in theory) entrepreneurship to distressed or growth-potential areas. Industrial employment problems can be addressed by encouraging workers in declining industries to develop a different set of skills or by creating jobs that demand skills individuals already have.

The perennial issue of jobs to people versus people to jobs (Hoover and Giarratani, 1984: 361) has been widely discussed in the urban and regional policy literature. Historically, U.S. policies have placed greater emphasis on attracting jobs to people. The Appalachian Regional Development Act, for instance, includes the phrase that jobs should be provided for people wherever they choose to live. Urban Development Action Grants (UDAGs) and the proposed Urban Enterprise Zones are examples of geographically oriented programs intended to attract job-creating capital to distressed areas. Efforts to move people to job-growth areas have been modest although some employment programs have included geographic relocation assistance. Proposals for more relocation efforts have been included in comprehensive urban policy proposals (U.S. Department of Housing and Urban Development, 1982).

The industrial policy perspective has emphasized the importance of creating job opportunities in industrial sectors with established work forces. For instance, Rohatyn, Shapiro, and Kirkland proposed an Industrial Finance Administration (Center for National Policy, 1984: 7–8) to channel capital to appropriate sectors. Efforts to increase the interindustry labor mobility include the Retraining Insurance Loan Fund proposed by Gary Hart. It would provide funds to retrain workers who have been laid off in declining sectors. Retraining is an industrial policy equivalent to the urban policy of moving people to job opportunity areas.

TARGETING

Among advocates of urban policy there has been a debate on whether assistance should be targeted toward expanding or declining cities. Although targeting assistance on the most needy cities has enjoyed support from a wide political spectrum, many critics of current urban policy have argued that efforts to prop up obsolete regions are wasteful and subsidies to growing areas will be more effective (President's Commission on Policies for the Eighties, 1981).

A similar division exists among industrial policy advocates: Should assistance be targeting toward declining or expanding industries? Labor leaders normally favor the former, resulting in the occasional industrial bail-out and tariff protections that support sunset industries. Contrasting views are held by the "Atari Democrats" who support channeling capital to emerging high-technology sectors.

In theory, the solution to the controversy is simple—target toward the sector that generates the most net new jobs per dollar of public subsidy. Unfortunately, the empirical answer to the question is unknown so any generalization along these lines will be incomplete. Concern about government's inability to identify emerging industries is a frequently raised objection to industrial policy in general (Krugman, 1983).

COORDINATING URBAN AND INDUSTRIAL APPROACHES

Urban and industrial programs can complement each other but they can also operate at cross-purposes if the programs are poorly designed. Howell (1980: 78) warned that "we could conceivably wind up with a highly successful industrial policy that contributes to further erosion of the investment base in cities." An example of a well-coordinated policy having both urban and industrial implications might be a program that simultaneously provided training in high-tech occupations in labor surplus areas and subsidized firms that employed high-tech workers if they locate in labor surplus areas.

Figure 16.1 represents a two-sector model that illustrates the importance of coordinating urban and industrial approaches. VMP_x and VMP_y represent the initial demand for labor (value of the marginal product) in a city. The VMP_x is drawn from right to left so L-D is the number of workers hired in industry X if the wage is w*. OD workers would be hired in industry Y. Assume that the initial wage is fixed at w* and cannot be lowered because of institutional factors. Further assume that the city is too small to have any affect on national wage or employment levels. Capital is perfectly mobile.

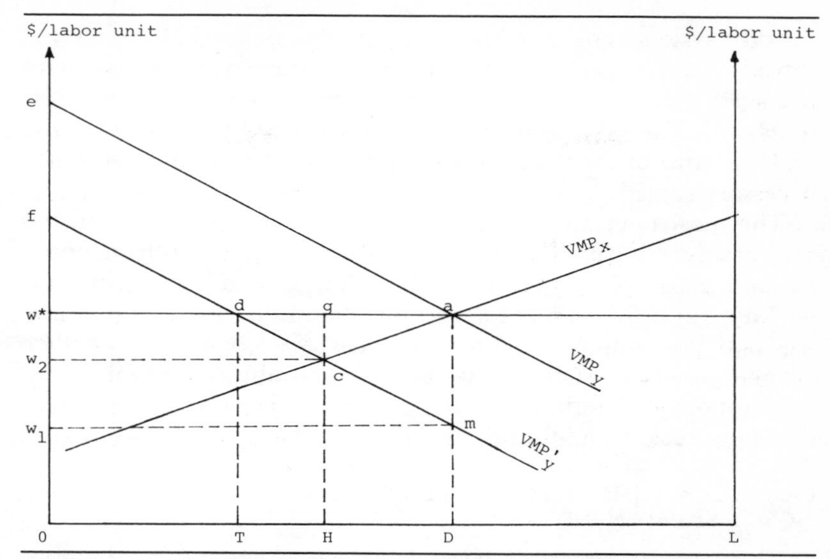

FIGURE 16.1: A Two-Sector Urban Employment Model

Initially the economy is shown in equilibrium with OD workers employed in industry X and DL workers in industry Y. Let some change—say, international competition or plant obsolescence—reduce the local demand for labor in industry Y to VMP'_y. Because wages are fixed at w*, the number of workers in Y declines to OT. Unemployment becomes TD. The consequences and the appropriate responses to the decline in labor demand depend upon the extent of labor mobility and wage flexibility. Three cases will illustrate the workings of the model.

CASE 1: LABOR IMMOBILE
BETWEEN INDUSTRIES AND REGION

This case combines the strongest assumptions that normally support both industrial and urban policy. The only way to (re)establish full employment is to subsidize local job creation in industry Y.

Because labor is immobile between industries, wages in sector Y would have to fall in w_1 in order to maintain full employment. However, the assumption of rigid downward wages results in unemployment of T-D. All of the unemployment is concentrated in Y.

Under these circumstances the government could initiate a tax subsidy scheme to (re)establish full employment. A variety of tax-subsidy systems could achieve full employment (that is, a wage subsidy of w^*-w_1 per workers in Y). However, the minimum subsidy needed to induce firms in industry Y to hire TD additional workers at w^* is dma. The subsidy is the difference between the wage rate, w^*, and the value of the marginal product. It is therefore the minimum necessary subsidy.

The most interesting policy conclusion that emerges from this special case is that employment maintenance requires both an industry and geographic-specific subsidy. Aid should be targeted on declining industries and in declining cities. Both urban and industrial policy development has been affected by a debate about whether "winners or losers" should be the target of subsidies. Given the strong assumptions of interindustry and interurban labor immobility, the declining sectors should be subsidized.

CASE 2: LABOR IMMOBILE BETWEEN REGIONS, MOBILE BETWEEN INDUSTRIES

This case is interesting because we suspect that labor is more mobile between industries than between regions. Thus, it is a case that reflects realistic relative mobility (Alperovitz and Faux, 1984: 149). The first case ignored industry X because it could not absorb any of the workers dislocated from Y (due to the assumption of labor immobility between sectors). This case allows us to incorporate the interindustry impacts ignored in the previous case.

Figure 16.1 can also be used to represent this situation. The shift in the demand for labor from VMP_y' will cause a decline in employment by TD. (The shift in VMP_y will be larger when capital is mobile than when it is not mobile.) If wages were flexible, equilibrium could be reestablished at w_2 with OH workers in Y and L-H workers in X. But, due to the assumed rigid wages, subsidies will be required to induce firms in industries X and Y to hire additional workers. The minimum subsidy of dca will be required. The subsidy of dcg will induce firms in Y to employ TH workers and the subsidy of gca will subsidize the employment of HD workers in X.

At H, the change in employment per dollar of subsidy is equal between sectors. The saving compared to case 1 due to the interindustry mobility of labor is cma or dam-dac.

The assumptions of case 2 lead to the conclusion that a cost-minimizing subsidy should stimulate employment in both sectors. Furthermore, there is no compelling relationship between the size of the initial decline and the extent to which the declining sector should be subsidized. It depends upon the responsiveness of the respective VMPs to subsidies. The optimum subsidy will ensure that the marginal cost of subsidizing additional jobs will be equal between sectors.

CASE 3: LABOR IMPERFECTLY MOBILE BETWEEN REGIONS, MOBILE BETWEEN INDUSTRIES

This case assumes that workers can quickly shift between industries because they require no retraining or they can easily obtain necessary retraining. The exogeneous decline in labor demand in industry Y will initially result in unemployment of T-D. The unemployed workers will leave the area, but because of imperfect mobility migration will not be immediate. In reality it may take years. The outmigration can be pictured as shrinkage of the labor force in Figure 16.1 from OL to OL-TD until points d and a coincide.

There are two inefficiencies from the national perspective in the *laissez faire* response. First, the unemployment during the adjustment period represents lost output. Second, the T-D workers who are unemployed may not be the most mobile members of the labor force. Seniority or other criteria could be used to determine who will lose their job. The most efficient solution should encourage the most mobile members of the labor force to relocate.

The policy dilemma in this case is to subsidize employment in a way that will maintain employment without killing off relocation incentives. If the incentives to relocate (whether push or pull) are the same for all members of the labor force, then the most mobile workers (those with the lowest relocation costs) will be the outmigrants.

The least cost method of (re)establishing or maintaining full employment would be to subsidize the employment of TH workers in industry Y at a cost of dgc and HD workers in industry X at a cost of gca. If the subsidy were financed by an appropriate tax on all workers in the area, incentives to relocate would remain. Individuals for whom the cost of relocation was less than the tax would move. Incentives to relocate would have the same effect.

IMPLICATIONS

The most immediate implication of the model is that partial regional mobility may require both urban-specific and industry-specific targets. This implication coincides with the empirical finding that neither industrial nor urban-oriented policies by themselves will be sufficient to provide sufficient employment growth. A dual approach will also minimize the employment maintenance subsidy.

The extent that programs should rely upon urban or industrial approaches should reflect cost of migration. For instance, if labor were geographically immobile but relatively mobile occupationally, whereas the job opportunities were geographically footloose but the occupational mix was fixed, then a program that subsidized firms to locate in distressed cities would be more preferable. If labor were relatively mobile geographically but occupationally immobile, whereas jobs could be created effectively only in a narrow range of occupations, then a policy to move people to jobs would be effective.

Conservatives who oppose both urban and industrial policy would argue that the price system provides not only the correct signal but also a sufficient inducement for the appropriate relocation of resources, both occupationally and geographically. For instance, if labor were relatively mobile occupationally, then the price system will provide incentives for job seekers to undertake the cost of changing occupations; if labor is relatively mobile geographically, then the market will provide incentives to move. Those who favor structural employment approaches generally believe that the market often fails to provide correct signals, that the signals are not received, that the costs of mobility are high, and the job creating involves substantial externalities.

A POLICY MATRIX

Three employment policy strategies have been discussed. Urban policy targets geographically, industrial policy targets industrially, and macro policy is neutral in its deliberate target. The comparison of urban and industrial employment policy indicated that both approaches (1) create job opportunities, (2) provide locational (in a regional or an industry) incentives, or (3) carry out both activities simultaneously. The two-sector model indicated that efficient structural adjustments may occur at the same time among regions and among industries.

How does the current mix of employment programs compare to the possible array of approaches? Table 16.3 is a policy matrix showing the possible combinations of approaches. The most highly structured package would be in the upper left-hand cell. A program that trained workers and created employment opportunities in a specific industry in a specific city and encouraged workers to relocate to that area would be an example of this combination. Perhaps the closest example of the highly structured approach is the support for the space industry that has been concentrated in a few southern and western cities. The space program is not intended to be an employment program, however.

There is probably room for disagreement regarding what cell a particular program fits. However, Table 16.3 is a useful way to view the three policy approaches. Industrially oriented programs may be classified as (1) creating or maintaining job opportunities in specific industries, (2) affecting the industry/occupation of workers, or (3) doing both simultaneously. Other employment programs are neutral industrially. Likewise, urban-oriented programs operate by (1) creating or maintaining jobs in cities, (2) affecting the geographic location of workers, or (3) doing both simultaneously.

The programs listed in Table 16.3 are not all pure employment programs because many have multiple purposes. For instance, Community Development Block Grants (CDBG) and tariffs and quotas include employment as one of several purposes. Furthermore, the matrix may have excluded some programs that have very indirect employment impacts. Highway programs, for instance, are excluded because their intended employment effects are too ephemeral to the program. FmHA is a program that creates jobs principally in agriculturally related activities in rural areas but also subsidizes other activities; it is targeted toward rural areas. It is obviously not an urban program, but is included as an example of this combination of industrial-geographic targeting. Fiscal and monetary policies are considered neutral regionally and industrially largely because of their basic motives, although their actual impact may not necessarily be neutral.

The program categorization is viewed from the federal perspective. A city may use enterprise zones to stimulate a particular industry although the federal government has no intention of stimulating a particular industry. Thus, the program is industrially neutral from a federal perspective but it could (and should?) be administered locally in a way that was industrially targeted.

One conclusion that can be drawn from Table 16.3 is that most programs are neutral in either the industrial or urban target. This result may be due to the fact that when industry groups develop and lobby for legislation they are not concerned with the urban implications. Likewise, when urban representatives develop programs they are neutral regarding the industrial impacts. Theoretically, this will make the programs less efficient if the free market adjustment

TABLE 16.3
The Program Mix

Urban Target	Industrial Target: *Simultaneous*	*Create (maintain) job opportunities*	*Affect industry/ Occupation "location" of workers*	*Neutral*
Simultaneous				NT**
Create (maintain) job opportunities		FmHA		CDBG UDAG EZ*
Affect geographic location	PE**		TAA JTPA	
Neutral	T&Qs	IDB* FHA/VA	NSF grants	Fiscal and monetary policy YOW*

NOTE: CDBG, Community Development Block Grants to cities that may be used for a variety of purposes including economic development (the amount of assistance depends upon need indicators; UDAG; Urban Development Action Grants to local governments for job creation and local tax base improvement; *IDB, Industrial Development Bank to channel capital industries through interest subsidies and loan guarantees; TAA, Trade Adjustment Act, to workers who have been laid off due to international competition, provides training and relocation assistance; JTPA, Job Training Partnership Act trains workers for jobs in particular skills needed in particular areas; *EZ, enterprise zones create areas within distressed cities where government regulations and taxes are reduced; NSF, National Science Foundation grants for advanced study and research in particular industries; T&Qs, Tariffs and Quotas to support prices and limit competition in particular industries; FHA/VA, Federal Housing Administration/Veterans Administration loans allow lower interest rates and lower downpayments for home purchases; FmHA, Farm Home Administration loans for farms, housing, and infrastructure in rural areas; *YOW, Youth Opportunity Wage is a lower minimum wage level for youth; **NT, new towns to attract individuals to particular locations and provide jobs in those areas; and **PE, CETA public service employment encourages employment of workers in public-sector jobs.

* Proposed program; ** expired programs.

requires labor mobility between both industries and regions but labor is immobile in both directions. For instance, channeling capital to targeted areas (for example, UDAG) could be more effective if the local industry with the most job creating potential could be targeted.

However, lacking an omnipotent governmental targeter, the current program mix has some advantages. A program that encourages job formation in a particular region leaves the choice of which industry to encourage to the discretion of other governmental units and to private investors. This may be an efficient way to coordinate between units of government and between the government and the market. Similarly, most assistance that is targeted toward industry is not region specific. Thus, tariffs may protect an industry, but the particular region that benefits is left to the interactions of the market and local policymakers through their implementation of city-specific programs.

NOTES

1. Urban and industrial policies have multiple goals of which employment is one. For instance, industrial policies may have national defense preparedness goals, and urban policies may contribute to better cities. However, in this chapter we are concerned only with the employment aspect of these policies.

2. Thanks to Dennis Cantrell for performing the data collection and analysis. The counties analyzed were as follows: Pennsylvania—Allegheny; Wisconsin—Milwaukee; Illinois—Cook; Massachusetts—Suffolk; Michigan—Wayne; New York—Erie, New York, Monroe; Ohio—Cuyahoga, Lucas, Montgomery, Summit; and New Jersey—Essex, Hudson.

3. The shift-share formula is:

$$\Delta e_i = e_i[\Delta US/US_i] + e_i[\Delta US_i/US_i - \Delta US/US] + e_i[\Delta e_i/e_i - \Delta US_i/US_i]$$

$$\Delta E = \Delta\Sigma^{e}{}_{i}$$

where

e_i = employment in industry i among the sample countries in 1978,

Δ = change in employment in between 1978–1982,

US = total U.S. employment in industry in 1978,

ΔUS_i = total U.S. employment in sector i in 1978, and

E = total employment in the urban county sample.

4. A variant of industrial policy suggests that programs should not attempt to increase growth in particular sectors but should attend to structural changes that will encourage aggregate growth in unspecified industries (Premus, 1985). Encouragement of basic research and other forms of risk taking are examples of this approach. A nontargeted, aggregate industrial policy could best be simulated by affecting the national share component.

REFERENCES

ALPEROVITZ G. and J. FAUX (1984) Rebuilding America. New York: Pantheon.

BLAIR, J. (1973) "A review of the filtering down theory." Urban Affairs Quarterly (March): 303–316.

BLUMENTHAL, S. (1983) "Drafting a democratic industrial plan," New York Times Magazine (August 28).

Center for National Policy (1984) Restoring American Competitiveness: Proposals for an Industrial Policy. Washington, DC: Author.

HOOVER, E. M. and F. GIARRATANI (1984) Introduction to Urban Economics. New York: Alfred Knopf.

HOWELL, J. (1980) "Testimony on urban revitalization and industry policy." Committee on Banking, Finance, and Urban Affairs, U.S. House of Representatives, 96th Congress, 2nd Session. Washington, DC: Government Printing Office.

JACOBS, J. (1984) Cities and the Wealth of Nations. New York: Random House.

JAMES, F. (1984) "Urban economic development: a zero sum game?" pp. 157-173 in R. Bingham and J. Blair (eds.) Urban Economic Development. Beverly Hills, CA: Sage.

JASINOWSKI, J. (1980) Testimony on Urban Revitalization and Industrial Policy. Washington, DC: Committee on Banking, Finance, and Urban Affairs, 96th Congress, 2nd Session.

Joint Economic Committee (1984) The 1984 Joint Economic Report. Washington, DC: Government Printing Office.

KRUGMAN, P. R. (1983) "Targeting industrial policies: theory and evidence." Federal Reserve Bank of Kansas City Symposium on Industrial Change and Public Policy, Kansas City Federal Reserve Bank, August 25–26.

LEVEN, C. (1978) The Mature Metropolis. Lexington, MA: Lexington Books.

MULLER, T. (1981) "Regional-urban policy: should the government intervene?" Challenge (March/April): 38-41.

Organization for Economic Development and Cooperation (1983) Positive Adjustment Policies: Managing Structural Change. Paris: Author.

PREMUS, R. (1985) Venture Capital and Innovation. Washington, DC: Joint Economic Committee.

——— and C. H. BRADFORD (1984) Industrial Policy Movement in the United States: Is it the Answer? Washington, DC: Government Printing Office.

President's Commission for a National Agenda for the Eighties (1982) Urban America in the Eighties. Washington, DC: Government Printing Office.

REICH, R. (1983) The Next American Frontiers. New York: Times Books.

SCHWARTZ, G. G. (1980) Urban Revitalization and Industrial Policy: The Issues in Urban Revitalization and Industrial Policy. Committee on Banking Finance and Urban Affairs, Washington, DC.

U.S. Bureau of the Census (1984) Statistical Abstract of the United States, 1983. Washington, DC: Government Printing Office.

U.S. Department of Housing and Urban Development (1982) The President's National Urban Policy Report, 1982. Washington, DC: Author.

17

The Politics of Revitalization: Public Subsidies and Private Interests

TERRY F. BUSS and F. STEVENS REDBURN

□FOR THE LAST decade, the core of national and local urban economic development policies has been the use of direct public subsidies for new private development and jobs in distressed areas. These policies, and a new array of quasi-public institutions and development programs, have been legitimized under the phrase—authored by the Carter Administration but accepted by the Reagan Administration as well—"public/private partnership." The philosophy of the "new" partnership has been summarized by the Committee for Economic Development (CED, 1982: 4):

> The economic base of an urban area provides goods and services for local consumption, jobs and income for residents, and a resource and tax base for community services and public amenities. Economic development today requires conscious effort by local government and private-sector organizations to retain existing businesses, facilitate the opening of new businesses, attract outside investment, create jobs for local residents (with special attention to training and placement of the hard-to-employ), and foster a climate supportive of business expansion that is compatible with other community goals.

According to this CED manifesto:

- Private initiatives include actions by businesses or business associations in development, assessment and planning, promoting the community's image, establishing favorable purchasing and hiring policies, providing technical assistance to other firms, and making corporate operating and investment decisions (especially with regard to siting) designed to enhance the local economy.

- Government initiatives include helping businesses to expand or solving problems that could lead to their departure, promoting the community to attract new investment, assembling land for development, providing physical infrastructure and public services, establishing incentives for business growth, and creating a healthy business climate.
- Collaborative efforts by government and private organizations include establishing joint development goals, joint-venture development projects, joint efforts to train and find jobs for the hard to employ, and "financial packaging" that includes a mix of resources that appropriately reflects the public and private interest in development."

In this chapter, analysis focuses on the use of direct public subsidies to assist private sector development (see Hamilton, et al., 1984). This use of public funds distinguishes current economic development policies from the more traditional indirect public sector contributions to economic development through the building and maintenance of public capital infrastructure, the creation of new serviced sites for industry, education, training, and other supportive public services.

PUBLIC SUBSIDY AND PRIVATE INTEREST: A CRITIQUE

The relatively new use of public authority to fund private enterprise directly has been increasingly attacked by market-oriented conservatives who view such spending as inherently harmful to the economy. The Office of Management and Budget (OMB) under David Stockman repeatedly attempted to eliminate the major federal business subsidy programs, including those aimed at urban and regional revitalization. Development subsidies are said to distort patterns of private investment, thereby reducing the national economy's growth and competitiveness. Typical of these arguments is that directed by conservatives against the Urban Development Action Grant (UDAG) program (Ferrara, 1985: 3).

> The grant program, created in 1977, was intended to foster economic improvement and job creating in distressed urban areas by stimulating private investment with federal funds. But UDAG does not create new net investment or jobs for the economy as a whole. It simply redistributes investment and jobs from one part of the economy to another.

> And in the process, the program results in a new drain on the economy as a whole and a loss in total jobs.
>
> UDAG now serves as a sort of "urban slush fund," redistributing investment and jobs toward politically powerful and influential cities at great costs to other jurisdictions.

Stockman, testifying before Congress in April 1985 also argued that UDAG's contributions to local economic growth "have to be weighed against the economy—wide harm to trade-competitiveness and agricultural sectors caused by high budget deficits—to which UDAG contributes." In his view, this harm "far exceeds" the value of any UDAG-induced local economic gains (Stockman, 1985).

Many theoretical criticisms of subsidy programs arise from a philosophy that attributes greater wisdom to the marketplace than to any conscious investment strategy pursued by the public sector. Defenders of such programs, conversely, point to the critical role that government has played in the development of agricultural regions and in shaping urban America through massive use of subsidies for transportation, housing, and the defense industry. They argue that a public role in regional development is not only potentially useful but necessary and inevitable. Nevertheless, the use of direct public subsidies for private ventures ranging from condominiums and restaurants to steel mills and office towers raises new practical as well as theoretical questions about the role of the public sector in economic development.

Even if theoretically defensible, subsidy programs must also be justified in terms of their actual performance. Direct use of subsidies may involve practical problems that are difficult to remedy either through better design or more careful administration. Subsidies are difficult to target to the places that most need them. It is difficult for public agencies to avoid subsidizing projects that would have gone forward anyway,with wholly private financing. Many of the intended benefits to poor people and distressed places leak away to affluent people and nondistressed places. And, many marginal projects fail financially, or at least fail to grow as expected.

THE TARGETING PROBLEM

Targeting is accomplished, at a national level, through the use of need formulas, such as the UDAG "distress" and "impaction" formulas used to establish city eligibility, and similar formulas used to rank projects in the competition for funding (Gatons and Britnall,

1984). However, the political compromises embodied in these formulas as well as serious data limitations produce a distribution of funds that is sometimes inconsistent with the programs' economic theory. Eligibility is usually defined in broad terms to ensnare congressional majority support. Further distortions are introduced by variations in local governments' aggressiveness and capacity to generate fundable project proposals. For instance, through the first five years of the UDAG program two of the most entrepreneurial cities—Baltimore and New York—each received over 50 UDAG awards' but 20% of the almost 400 UDAG-eligible large cities and urban counties received no awards (Gatons and Britnall, 1984).

UNWARRANTED SUBSIDIES

One of the nation's largest corporations, a member of the *Fortune* 500, acquired an abandoned industrial site adjacent to its headquarters—a site centrally located in one of the fastest growing and most affluent communities in that state. The corporation possessed enormous capital and was moving into new markets. It received debt interest subsidies from the public sector for expansion on this site, using the arguments that it could not otherwise afford the investment and that it might move its facilities. This particular expansion displaces local medical services that are in high demand and that cannot be transferred out of the community.

It is difficult for public officials to know the true intentions of private companies or to determine whether public subsidies are necessary to make a particular project feasible and sufficiently profitable. The near impossibility of distinguishing between situations in which subsidies are needed and those in which they merely produce profit windfalls draws local public officials into destructive and costly bidding wars for new investment.

BENEFITING THOSE LESS IN NEED

The benefits of subsidy programs tend to be distributed almost as widely as the benefits of any nonsubsidized private investment. Thus, the lion's share of profits goes to shareholders and lenders who pay taxes far from the distressed city where the project occurs. If a project creates fine shops, luxury housing, or corporate office space for managers and engineers, it may also create entry-level jobs, new revenues for financially pressed jurisdictions, and new customers for small store owners. Nevertheless, projects such as the UDAG to Stowe, Vermont, and the UDAG to Newport, Rhode Island, for hotel construction in these two upper-class resort towns invite skepticism regarding the balance of benefits between poorer and more affluent people.

SUBSIDIZING MARGINAL ENTERPRISES

One distressed city had attracted few industrial prospects, in spite of extensive promotional activities and the creation of a heavily subsidized industrial park. Then the city was approached by a firm that promised to create many jobs in the area. City officials put together an impressive local, state, and federal development finance package. In their eagerness the local officials required little in the way of formal agreements with the firm and never appreciated the fact that there was no market for its products. The firm soon failed and the development incentives were "eaten up" by the private entrepreneurs. In addition, the firm removed equipment and materials valued at hundreds of thousands of dollars from an acquired industrial building and sold them at a profit. Because of the way the subsidies were expanded and the on-site sales were managed, the city was forced to sue the firm. After three years, it has recovered nothing. Further, the city now cannot lease a large proportion of its industrial sites because of restraining orders by the courts. The litigation continues.

No one has as yet looked systematically at the frequency with which blue-sky and blue-suede-shoe operations are drawn to the honey pot of subsidy programs. No one has estimated the proportion of subsidized projects that quickly fail or the resulting cumulative waste of public funds.

DISPLACEMENT OF MORE DESIRABLE INVESTMENT

Even more fundamental than these problems with the performance of subsidy programs may be the opportunity costs associated with such efforts. The kinds of projects subsidized are not always the kinds of projects that the distressed area or city most needs. In some cases, the resulting private investment may actually displace other, more desirable kinds of private investment—either those that would require subsidies or those that would not. In other cases, the public money spent on subsidies might have done more to foster economic development if spent on more traditional government responsibilities such as education or the provision of services or infrastructure.

The problems of the U.S. steel industry provide an illustration. Domestic production of basic steel is in considerable economic trouble: high production costs, foreign competition, dwindling markets, environmental regulation, and other factors are slowly wreaking havoc in this industry. One city has poured subsidies into the operations of two steel mills, both of which have failed. A new subsidized steel facility of the same kind is now under negotiation. In order to attract this latter venture, the city is required to create an enterprise zone, forgive all property taxes, and provide loans and interest subsidies. By working such a deal, the city has denied itself much needed tax revenues and greatly reduced its capacity to assist other more viable operations, all in a desperate effort to hold onto its share of a failing industry.

EXPLAINING THE MISUSE AND ABUSE OF SUBSIDY PROGRAMS

Accumulating experience with the implementation of subsidy programs at the local level suggests several factors that contribute to their misuse. These factors include:

- the lack of a strategic planning capacity;
- private sector dominance of subsidy programs;
- mismatch of private interests and public responsibilities;
- the rise of public entrepreneurs; and
- off-budget decision making.

LACK OF STRATEGIC PLANNING CAPACITY

The lack of expertise in strategic development planning in local government contributes to the careless and wasteful use of subsidies. Strategic planning

> requires the explication of alternatives; forces a future orientation; encourages broad-scale information gathering and evaluation; provides for communication and participation; emphasized the importance of implementation. . . [Olsen and Eadie, 1982: 7].

Analysts have recognized that the absence of strategic planning capacity in the local government generally has hampered public policy programs and initiatives in at least five ways (Walters and Choate, 1984: 11–12):

(1) Fragmented and disordered management practices permit no overall view of public needs and no overall specification of the roles of the respective levels of government or the private sector in meeting those needs.
(2) In the absence of clearly documented needs and well-articulated priorities, pork-barrel politics often dominates public actions and public expenditures.
(3) In this atmosphere, the short-term payoff is invariably favored over long-term goals. When public expenditures are treated as a source of patronage, it becomes difficult to create and maintain coalitions that can sustain long-term efforts, no matter how vital such efforts may be. As a result, most political leaders support those programs that can be financed and completed in a single year.
(4) Without coherent strategies that set investment needs and ways to meet them, specific public investment plans cannot be systematically formulated. Nor can government leaders and the public determine critical linkages between various public functions.
(5) Disordered management practices at one level of government are easily transferred to other levels.

In the past, the lack of strategic planning capabilities may have made little difference in local economies: The growth of many large cities prior to the 1970s occurred in spite of or without the the need for local public planning or strategic interventions. However, in the 1970s and 1980s planners have faced the dual challenge of stemming the tide of precipitous decline of established industries, on the one hand, and the need to position local economies so as to maximize growth opportunities should any arise, on the other hand. Public subsidies for development offer what appears to be a quick-fix solution, which is short term and highly visible. The absence of strategic thinking and analysis make the temptation that much harder to resist.

PRIVATE SECTOR DOMINANCE OF SUBSIDY PROGRAMS

Local governments, always vulnerable to the claims of private capital, are especially so in hard times.

Local capital interests, including landowners and commercial or industrial developers, typically exert strong, direct leverage on public subsidy decisions. These economic actors benefit from subsidies in various ways:

- less profitable or unprofitable vacant or unused land or facilities owned or controlled by a developer can be made to yield a profit;
- land or facilities owned by others can be made profitable, indirectly benefiting the developer's existing investments or interests;
- losses by a developer can be recouped or prevented by engaging public participation; or
- competitors of a developer can be stymied or held at bay by marshaling public support.

Developers are often able to control the economic development decision process for several reasons. First, economic development is extremely complex, requiring detailed knowledge and experience in management, legal affairs, financing, construction, and land economics in order to be successful. Few actors in the marketplace or political arena have or can afford to purchase this expertise. Second, developers tend to be well integrated into both public and private sector networks that make decisions about development. Developers, or more likely their surrogates, are often members of public planning or zoning boards and commissions, local economic development corporations (EDCs), Private Industry Councils (PICs), or economic development task forces appointed by mayors, county commissioners, or Chambers of Commerce. Directly or indirectly, developers may be either catalysts or initiators of development, whereas most others are relegated to the sidelines. Third, developers, out of necessity, are often politically active, especially in influencing candidates for or holders of public office. Campaign contributions and other forms of political influence are commonplace. Because of this, developers have a considerable say in what the public sector does or does not do. Finally, developers have a considerable stake in the community and are likely to have been and continue to be involved in the community over time. Having much to lose and to gain, local developers and landowners are willing to fight the battles necessary to dominate the revitalization process.

PRIVATE OPPORTUNITIES VERSUS PUBLIC RESPONSIBILITY

The misapplication of public business subsidies has also to do with the inherent difference between the incentives/opportunities for private investors and the public responsibilities that public officials are charged with. The long-term development interest of a locality and the short-term profit-seeking considerations that drive private developers and investors should often lead the two sectors to different conclusions regarding what kinds of development are most desirable. This is especially so in declining areas, where the public sector should be most aggressive in pursuing an adaptive economic development strategy.

Nearly all private entrepreneurs—at least those who have survived over time—invest their capital with the expectation that the return on investment will be commensurate with the risk undertaken: the greater the risk, the greater the expected return. Reinvestment has declined and disinvestment has accelerated in distressed cities because private entrepreneurs perceive returns on investments to be insufficient and attendant risks to be too high.

The public sector, on the other hand, has traditionally and properly invested its capital in areas where the financial return is long term, indirect, and uncertain. Public support for infrastructure improvements and education are prime examples of this traditional public role in support of local economy, which complements and supports investments made by the private sector. With the rise of subsidies, the sharp distinction between public and private objectives has blurred. Because the public sector is often the weaker partner in negotiations with the private sector—for reasons discussed—it is vulnerable to arguments that any new private investment advances the goal of revitalization. The best way for the public sector to escape from its vulnerable position may be to back out of the subsidy game and return to more traditional methods of supporting economic development. However, many in the public sector now enjoy and benefit personally from the subsidy programs.

RISE OF PUBLIC ENTREPRENEURS

Subsidies may be misapplied more frequently today due to the rise of a new bureaucratic class of "public entrepreneurs" whose career interests, thinking processes, and, often, backgrounds are like those of the private sector developers and investment bankers with whom they deal. For present purposes, a public entrepreneur may be thought of as "one who gathers and risks political capital or support in order to reshape politics and create new sources of power by establishing new programs (or products)" (Mollenkopf, 1983: 6).

Although revitalization can be viewed purely as a question of economics, revitalization is at the same time a way of using public resources to further the interests of particular individuals in both public and private sectors, thereby serving as a political resource for public entrepreneurs.

This entrepreneurial approach, and the rise of this new class, has been promoted through HUD's administration of the UDAG program. In bargaining with developers cities are encouraged to maximize their net financial return from each project through inclusion of loan payback and profit participation provisions in UDAG contracts. HUD's emphasis on this form of benefit reduces the city's ability to negotiate with private investors for other benefits. This makes it harder to use UDAG for projects with less direct financial return but more in line with cities' long-run development needs (Clarke and Rich, 1985).

"OFF-BUDGET" DECISION MAKING

The tendencies described above are all reinforced by the movement, at all levels of government, to off-budget investment mechanisms that take subsidy decisions out of the political limelight and, more and more frequently, place them under direct private sector control (Clarke and Redburn, 1983).

These off-budget mechanisms include any continuing publicly created incentives to the private sector that do not require specific legislative authorization or annual appropriations. The two major categories are "tax expenditures," such as the federal income tax exemption for interest on industrial revenue bonds (IRBs), and "credit activities," such as publicly capitalized revolving loan or venture capital funds and public loan guarantees.

Both public and private sector entrepreneurs gain by using off-budget, rather than on-budget, development instruments. For instance, when IRBs are awarded by a quasi-public development corporation and approved by local political officials, typically no public hearings are held to determine whether or not issuance is in the public interest. Officials may take credit for development, or quietly allow deals to occur without any public notice. If a quasi-public corporation awards the bonds, only the public sector is even less directly accountable. In the unlikely event that a public outcry occurs, officials may argue that no one in the local area is hurt by the issuance of bonds—the only loser is the U.S. Treasury.

Private-sector entrepreneurs thrive on the subsidies that off-budget devices allow; and private interests can more readily influence the use of public funds when there is no direct appropriation or other opportunity for public debate prior to the decision to invest.

CONCLUSION

The greatly expanded use of direct public subsidies for private development during and after the Carter years has allowed the symbolism of economic development to justify policies that, not surprisingly, serve the interests of local capital and political entrepreneurs. Unfortunately, the politics of revitalization and the recent reliance on direct subsidies have done little to eliminate fundamental barriers to the economic recovery of distressed urban areas. These areas need to develop a new social/institutional base that will support human development and investment, while fostering the political support and governing capacity to sustain recovery. There are alternatives, the briefest outline of which is offered below.

Soundly conceived local economic development policy must embody the following two principles, which are in conflict with the recent trend in U.S. economic development policy:

(1) *Development is not equivalent to private investment but depends, rather, on particular kinds of private investment.*
(2) *The right kinds of private investment will only be made if the requisite institutional and infrastructural base for such investments is present.*

The right kinds of private investment are those that integrate well with the existing local economy—or more specifically, that extensively use products and services of existing local industries in their own production, produce products and services that substitute for imports from outside the region, and generate exports to the outside. Preferably, also, the profits and personnel savings generated by these investments will be retained by local people and reinvested in the local economy. Such investments reinforce the existing economic strengths of an urban area and can set off a rapid chain reaction of related private investment. This, according to Jane Jacobs (1984) is the *only* way in which city economies grow.

Although direct subsidies used very selectively may support the right kinds of private investment, they are not the key ingredient in successful local economic development. Especially where the problem is revitalization of an area declining due to the loss of established economic functions, the primary public sector responsibility is for the institutions and infrastructure on which appropriate kinds of private investment depend.

The problems of infrastructural provision and institutional development are familiar in the context of nonindustrialized regions. Much less is known about the problems of reorienting and rebuilding these bases for growth in distressed urban areas. Incubation strategies approach the problem from a microeconomic perspective by starting with a fledgling firm and its entrepreneur, identifying their particular needs, and acting to meet their needs. From the rapidly accumulating local government experience with incubation strategies may emerge general knowledge about the needs of emerging firms, which can lead to other economic development policies focused on institutions and infrastructure. For instance, local government may work with local financial institutions to improve access to credit by new or expanding firms producing for local industry, or may work with local technical training and education institutions to provide specialized skills training to new or expanding firms. Such programs offer a possible route of escape from the subsidy trap into which all levels of U.S. government have lately fallen.

REFERENCES

CLARKE, S. and F. S. REDBURN (1983, October) "Off-budget urban policy." Presented at the Annual Conference of the Association for Public Policy Analysis and Management, Philadelphia.

CLARKE, S. E. and M. J. RICH (1985) "Making money work: the new urban policy arena," pp. 101-115 in Research in Urban Policy, Vol. 1. Greenwich, CT: JAI.

Committee for Economic Development (1982) Public-Private Partnership. New York: Author.

FERRARA, P. J. (1985) "It's time to end the urban slush fund." Washington Times (April 2), p. 30.

GATONS, P. K. and M. BRITNALL (1984) "Competitive grants: the UDAG approach," in R. Bingham and J. Blair (eds.) Urban Economic Development, Vol. 27, Urban Affairs Annual Reviews. Beverly Hills, CA: Sage.

HAMILTON, W., L. LEDEBUR, and D. MATZ (1984) Industrial Incentives: Public Promotion of Private Enterprise. Washington, DC: Aslan Press.

JACOBS, J. (1984) Cities and the Wealth of Nations. New York: Random House.

MOLLENKOPF, J. H. (1983) The Contested City. Princeton, NJ: Princeton University Press.

OLSEN, J. B. and D. C. EADIE (1982) The Game Plan: Governance with Foresight. Washington, DC: Council of State Planning Agencies.

STOCKMAN, D. A. (1985) Testimony before the Housing Subcommittee of the Senate Banking, Housing, and Urban Affairs Committee, April 15. (mimeo).

WALTER, S. and P. CHOATE (1984) Thinking Strategically: A Primer for Public Leaders. Washington, DC: Council of State Planning Agencies.

18

The Revitalization of New England Cities

JAMES M. HOWELL

□OVER THE PAST several years a great deal has been said and written about the economic revitalization of New England (Avault, 1983; Browne, 1984; Council for Economic Action, 1983; Harrison, 1984; Howell, 1984a, 1984b, 1984c, 1985; Howell and Frankel, 1983; Stevens, 1983). Indeed, the turnabout in the regional economy's industrial base from a state of near-stagnation as recently as the mid-1970s has been singularly spectacular—so much so that economic development officials and policymakers alike have beaten a path to the region's cities in the hope of discovering the secrets underlying this phenomenon.

There are no secrets behind the region's dynamic growth process. The growth has been driven by the highly interactive roles of five factors—higher education, research and development, entrepreneurship, capital, and labor—each of which played a role in the solid-state computer revolution. The birth of the minicomputer industry in New England has given the region an industrial comparative advantage that is not likely to be replicated elsewhere. However, other developments in the region's economy are also important. Most noteworthy is the increased economic impact of communities that have become regional trade and service centers. For decades the urban growth of New England was dominated by the Boston-New York axis. Now regional centers such as Portland, Worcester, and Burlington are becoming major contributors to the economy of New England.

AUTHOR'S NOTE: *I wish to express my appreciation to Ms. Diane Fulman, Consultant to the Economics Department.*

Yet, as one travels through the region, it becomes apparent that although some of the older industrialized centers are now enjoying a relatively high level of prosperity, others are not. The close juxtaposition of successful and not-so-successful cities (for example Nashua and Fitchburg, and Lowell and Lawrence) raises most interesting issues about the manner in which economic growth spreads throughout a region, and about the barriers that seem to prevent its benefits from being shared equally among all cities. These issues suggest a central question: How does a city "start over" industrially, and what factors promote or inhibit that new beginning?

A region as small as New England provides an excellent place to address this issue of "starting over," as the region's mill towns share many similarities. They are close to each other, both geographically and chronologically, because they all began their industrial growth at approximately the same time in the early- to mid-nineteenth century. Although their industry mix was initially diverse—ranging from textiles and apparel to shoes and machinery—it was generally labor intensive. Finally, by the early 1950s, all had begun to suffer from serious competition from the American South and foreign countries.

Seen from a distance the recent history of the region's medium-sized cities—the so called "old mill towns"—appears to have been one of steady progress under the impetus of New England's widely acclaimed high-tech growth. The results of our analysis, however, reveal the dramatic unevenness of the region's technological revitalization across cities, suburbs, and rural areas. In fact, we have found that only about one in four of the region's cities has been able to start over industrially by capitalizing on the growth of high-tech industries. The rest have been forced into other channels—or, sadly enough, left behind.

AN ANALYSIS OF SAMPLE CITIES

The New England economic landscape contains 111 cities, representing 43% of the region's total population and 43% of aggregate employment. In order to make our analysis manageable, we selected 35 medium-sized cities for detailed investigation, and concentrated our attention on the disaggregated economic performance data for each individual city. We selected the cities because they were generally representative of all cities within the region, and because we had first-hand familiarity with their attempts at starting over. We excluded Boston because of its very special circumstances.

TABLE 18.1
35 Sample Cities

	1970 (%)	*1980* (%)
Population shares as a percentage of		
Region	22	20
111 cities	45	45
Employment shares as a percentage of		
Region	22	19
111 cities	45	43
Manufacturing employement shares as a percentage of		
Region	25	22
111 cities	51	50

SOURCE: U.S. Bureau of the Census (1970, 1980) for the following states: CT, ME, MA, NH, RI, and VT.

Taken as a whole, the 35 cities selected for analysis represent roughly half of the aggregate economic activity of the region's 111 cities (see Table 18.1). Taken individually, the data for these sample cities (see Table 18.2) reveal three important characteristics:

- In 1970 all 35 cities, at least in terms of their per capita incomes and unemployment rates, were quite similar. The differences that show through are between the two subregions: the three northern states (Maine, New Hampshire, and Vermont) and the three southern states (Massachusetts, Connecticut, and Rhode Island). Yet even across these subregions, the statistical conformity among cities is remarkable.
- The comparatively low unemployment rates in the majority of the cities reflect the extent to which these cities benefited from the rapid national economic growth of the 1960s. They also reflect the geographical compactness of the area and the ability of previously displaced mill workers to commute to nearby suburban growth areas, especially along Massachusetts Route 128 and within other suburban growth nodes.
- Each of these 35 cities played a role in the initial industrialization process in New England. Although their industry mix varied somewhat (see the last column of Table 18.2 for the historically dominant indus-

TABLE 18.2
General Economic Characteristics for Selected New England Mill Towns, 1970

Central City	*Percentage Employment In Manufacturing* (%)	*Income Per Capita* ($)	*Unemployment Rate* (%)	*Traditional Industry Base*
Connecticut				
Bridgeport	43	3,233	4.7	Metal working
Bristol	50	3,544	4.1	Clocks, watches, fabricated metals
Hartford	26	3,113	4.5	Woolen mills
New Britain	47	3,509	4.4	Builders' hardware
New Haven	26	3,181	4.5	Guns, clocks, hardware
New London	26	3,428	4.1	Shipbuilding, naval equip., metal products
Norwich	33	3,108	4.1	Textiles
Waterbury	43	3,296	5.7	Brass milling
Massachusetts				
Brockton	30	3,074	4.1	Shoes
Fall River	45	2,677	5.2	Needle trades
Fitchburg	43	2,992	4.7	Paper
Lawrence	45	3,280	4.5	Textiles
Leominister	49	3,324	3.5	Wool carding, leather, shoes
Haverhill	43	3,324	5.0	Shoes
Lowell	37	2,867	4.4	Textiles
New Bedford	46	2,694	5.4	Textiles
Pittsfield	40	3,347	4.4	Wool, paper, shoes

Springfield	30	2,982	4.5	Fabricated metal goods, paper products, chemicals
Chicopee	46	2,965	4.5	Textiles, metal working
Holyoke	39	2,933	5.3	Textiles, metal mills
Worcester	30	3,242	3.9	Metal working, textiles, paper goods
Rhode Island				
Pawtucket	44	2,977	4.2	Textiles
Providence	33	3,110	4.4	Textiles, jewelry
Vermont				
Burlington	18	2,981	4.3	Metal working, food products
Springfield	48	3,211	4.5	Machine tools
Rutland	21	2,828	4.1	Marble, railroads, stone working machinery
New Hampshire				
Claremont	46	2,887	4.6	Mining, mill machinery, textiles, paper, shoes
Manchester	34	2,953	3.4	Textiles
Nashua	47	3,299	2.7	Textiles, machine tools
Portsmouth	27	2,746	3.7	Shipping, navy yard, paper goods
Rochester	53	2,703	7.1	Shoes, textiles, wood products, paper
Maine				
Auburn	40	2,826	4.8	Shoes
Bangor	14	2,553	4.1	Shipping, shoes, paper, lumber, wood products
Lewiston	42	2,543	4.2	Textiles, shoes
Portland	17	2,817	3.9	Food processing, lumber, wood products

SOURCES: U.S. Bureau of the Census (1970) for the following states: CT, ME, MA, NH, RI, and VT; Encyclopedia Americana, International Edition (1973).

tries) most have had and continue to have high ratios of manufacturing employment: The mean manufacturing ratio, at 37%, for the 35 sample cities is more than 50% above the national average.

The economic development of these 35 cities—having followed a path of early industrialization and having maintained a high ratio of manufacturing employment—is similar to that of many other older New England mill towns. By 1970, their economic plight had brought them all to more or less the same economic and industrial point—ready to begin the complicated process of "starting over."

STARTING OVER: THE REGIONAL AGGREGATES

What happened to these 35 cities during the decade of the 1970s? The answers are most significantly influenced by two crucial periods: 1968–1975 and 1975–1980 (see Table 18.3). The years 1968–1972 were years of transition for New England. The 1969–1970 national recession was a factor. In addition, because of the region's disproportionately high share of aerospace and defense-related industries, the sharp cuts in NASA and in defense spending during these years were deeply felt. Moreover, the high-technology sector—particularly the minicomputer industry—had not yet developed to its full potential. Finally, the special problems of excessive regional dependence on imported oil for energy became readily apparent after the first Arab oil embargo in the fall of 1973. Soon thereafter the regional economy was pulled into a deep recession. As the statistics vividly reveal, the region's manufacturing sector deteriorated sharply with many of the traditional manufacturing industries rendered less cost-competitive because of high energy costs. The negative weight of much of this adjustment was borne by the region's 111 older cities. In many respects, this new economic adjustment was more severe than the mill-liquidation period of the early 1950s.

Yet as the data also show, the region experienced a significant turnabout in the national business cycle growth period from 1975 to 1980. It was during this period that the considerable employment-generation aspects of the high-tech industry became most evident, producing the striking turnabout in the manufacturing sector. In Massachusetts alone the unemployment rate declined from over 12% in 1974 to below 4% in 1984.

TABLE 18.3
Percentage Change in Key New England Economic Variables

	1968–1975 (%)	*1975–1980* (%)
Total Employment	+5.2	+17.5
Construction	-0.5	+ 0.3
Manufacturing	-5.6	+ 4.4
Trade	+3.0	+ 3.7
Services	+5.2	+ 7.3
Government	+3.1	+ 1.8

SOURCE: Employment and Earnings, Bureau of Labor Statistics, U.S. Department of Labor.

STARTING OVER: THE RESULTS

In large part, this revitalization process affected suburban areas and the smaller towns. Some benefits did spill over into the cities, but the effects were by no means uniform. In fact, an analysis of the data suggests that the 35 sample cities can be conveniently grouped into several broad categories, depending on the extent of their participation in the high-technology revitalization process. Although we recognize that there are limits to any scheme of categorization, our examination of dozens of variables revealed the emergence of some general patterns across the 35 cities. These patterns allow us to differentiate between cities that have started over and those that have not by dividing them into three groups: international high-technology centers, industrially mature mill towns, and regional service centers.

- *International high-tech centers.* The cities that appear to have been significant economic gainers as a result of the high-tech revitalization process during the latter half of the 1970s are listed in Table 18.4. Because of data constraints, we have defined the "high-tech sector" as the three industries with Standard Industrial Code (SIC) numbers 35 (nonelectrical machinery), 36 (electrical machinery), and 37 (transportation equipment). Without question, the most significant conclusion to be drawn from this research is that only 9 of the 35 cities analyzed fall into this category. Of the 9, 3 are in New Hampshire. These statistics confirm the view that the high-tech revolution in New England has primarily benefited suburban and rural communities rather than older industrialized cities.

TABLE 18.4
Cities Revitalizing as International High-Technology Centers

	Percentage Employment in High-Tech Industries (%)	*Percentage Increase in High-Tech Employment 1970–1980* (%)	*Increase in Per Capita Income 1970–1980* (%)	*Unemployment Rate 1980* (%)	*Percentage Point Change in Unemployment Rate 1970–1980* (%)	*Percentage of Families Living in Poverty*		*Minority Population 1980* (%)
						1969 (%)	*1979* (%)	
Connecticut								
Norwich	15	+67	+114	6.6	+2.5	8.1	10.3	6.1
New London	17	+39	+88	7.4	+3.3	10.2	13.0	22.1
Bristol	25	+18	+118	4.8	+0.7	3.7	4.7	3.7
Rhode Island — None								
Massachusetts								
Lowell	21	+89	+110	4.8	+0.4	8.5	11.3	6.8
Haverhill	21	+34	+94	5.9	+0.9	6.5	9.0	3.5
Leominster	11	+69	+106	4.3	+0.8	5.8	7.9	5.6
New Hampshire								
Portsmouth	18	+59	+134	4.6	+0.9	9.2	6.8	6.7
Rochester	16	+35	+133	5.8	-1.8	7.2	5.9	0.7
Nashua	24	+95	+138	3.5	+1.8	4.8	5.4	3.0
Vermont — None								
Maine — None								

SOURCES: U.S. Bureau of the Census (1970, 1980) for the following states: CT, MA, NH, and RI; U.S. Bureau of the Census, General Social and Economic Characteristics, United States Summary (1980).

- *Industrially mature mill towns.* A total of 20 of the cities analyzed were found to have economic situations that were little improved, and in some cases actually worse, at the end of the decade (see Table 18.5). This category of old mill towns appears to have been left behind in New England's economic revitalization process. They now must struggle with the problems of industrial maturity and declining competitiveness.
- *Regional service centers.* The economic dynamics of 6 of these cities did not easily fit into either of the two above categories (see Table 18.6). A careful review of the statistics, however, suggests that they are emerging as regional service centers with high employment ratios in the nonmanufacturing sector. This development has been welcomed by some as an important step in breaking away from manufacturing, but it is not necessarily beneficial to the city itself. The 3 southern New England cities in this category are increasingly beginning to take on the unenviable characteristics of much larger old industrialized cities, with high ratios of unemployment, many families living in poverty, and large minority populations. Table 18.7 shows clearly the comparative socioeconomic performance of these 6 cities and their much larger regional counterpart—Boston.

If we leave aside these regional service centers—partly because they are few in number, and partly because there are differences between southern New England and northern New England centers—we can draw some useful distinctions between the remaining high-tech cities and the mill towns. As Table 18.8 shows, the high-tech cities far outshine the mill towns in three significant areas: unemployment (rising at only about half the averages of the United States and of the mill towns during the decade); manufacturing employment (dominated by high-tech systems and components); and total employment. In addition, as Table 18.9 shows, the high-tech cities experienced more significant increases in per capita income and a slower rise in poverty than did the mill towns.

THE OUTLOOK FOR OLDER NEW ENGLAND CITIES

To the extent that the experiences of these 35 cities are representa tive of the 111 cities in New England, the message about the comeback of Snow Belt cities as a result of high technology seems to have been exaggerated. If the large majority of New England's mill towns have been left out of the high-tech process, where will their economic future be found?

TABLE 18.5
Industrially Mature Mill Towns

	Percentage Employment in Traditional Manufacturing (%)	*Change in Manufacturing Employment 1970–1980* (%)	*Increase in Per Capita Income 1970–1980* (%)
Connecticut			
Waterbury	41	-4	+95
Bridgeport	40	-16	+88
New Haven	23	-20	+83
New Britain	40	-17	+104
Rhode Island			
Pawtucket	44	—	+113
Providence	32	-15	+98
Massachusetts			
Fall River	46	—	+94
Lawrence	44	-21	+72
Pittsfield	31	-22	+112
Chicopee	40	-8	+115
Holyoke	30	-33	+109
Brockton	25	-11	+100
New Bedford	46	-9	+102
Fitchburg	36	-17	+103
New Hampshire			
Claremont	43	+12	+124
Vermont			
Springfield	45	+5	+121
Rutland	18	+5	+123
Maine			
Lewiston	14	-11	+120
Bangor	12	-1	+142
Auburn	30	-21	+112

SOURCES: U.S. Bureau of the Census (1970, 1980) for the following states: CT, ME, MA, NH, RI, and VT; U.S. Bureau of the Census, General Social and Economic Characteristics, United States Summary (1980).

Unemployment Rate 1980 (%)	*Percentage Point Change in Unemployment Rate 1970–1980* (%)	*Percentage of Families Living in Poverty* *1969* (%)	*1979* (%)	*Minority Population 1980* (%)
6.7	+1.0	7.1	11.6	22.0
7.1	+2.4	8.6	17.4	40.2
9.3	+4.8	12.9	19.4	40.9
5.1	+0.7	6.0	8.9	15.1
7.0	+2.8	9.0	13.4	6.1
9.2	+4.8	13.3	20.5	20.5
7.6	+2.4	10.8	13.1	3.6
7.0	+2.5	8.7	17.0	18.1
8.1	+3.7	5.6	8.7	3.8
5.7	+1.2	5.4	7.7	2.2
7.0	+1.7	10.6	16.6	16.7
7.2	+3.1	6.5	7.7	8.5
9.0	+3.6	11.9	14.1	12.7
6.2	+1.5	7.7	9.2	4.6
3.7	-0.9	6.8	9.1	—
3.4	-1.2	6.3	7.7	—
6.7	+2.6	7.7	8.1	—
7.1	+2.9	8.7	10.0	1.1
8.3	+4.2	11.6	9.8	2.1
5.8	+1.0	7.6	9.9	0.8

TABLE 18.6
Cities Emerging as Regional Service Centers

	Percentage Employment in Nonmanufacturing Sectors 1980 (%)	*Increase in Per Capita Income 1970-1980* (%)	*Unemployment Rate 1980* (%)	*Percentage Point Change in Unemployment Rate 1970-1980* (%)	*Percentage of Families Living in Poverty*		*Minority Population 1980* (%)
					1969 (%)	*1979* (%)	
Connecticut							
Hartford	76	+79	7.7	+2.2	12.6	22.5	54.7
Rhode Island							
None							
Massachusetts							
Springfield	74	+95	7.2	+2.4	9.6	15.6	26.1
Worcester	73	+99	5.6	+1.7	7.1	11.2	7.6
New Hampshire							
Manchester	72	+132	5.2	+1.3	7.2	7.5	2.0
Vermont							
Burlington	83	+106	6.0	-1.2	7.8	7.8	2.1
Maine							
Portland	87	+128	6.4	+2.5	10.7	10.2	2.1

SOURCES: U.S. Bureau of the Census (1970, 1980) for the following states: CT, ME, MA, NH, and VT; U.S. Bureau of the Census, General Social and Economic Characteristics, United States Summary (1980).

TABLE 18.7
Comparative Analysis of 6 Emerging Regional Service Centers and Boston

	Percentage Change in the Decade of 1970s			
Six Regional Centers	*Nonmfg. Employment* (%)	*Families Living in Poverty* (%)	*Per Capita Income* (%)	*1980 Minority Population* (%)
Southern New England	-4.2	+68.3	+91.2	28.1
Northern New England	+20.9	-0.4	+121.7	2.1
Boston	-0.2	+42.7	+111.9	31.7

TABLE 18.8
Changes in Aggregate Employment, Manufacturing Employment, and Unemployment Rates in High-Tech Revitalized Cities and Industrially Mature Mill Towns

	Percentage Changes in Decade of 1970s		
Cities	*Total Employment*	*Manufacturing Employment*	*Unemployment Rate**
High tech	+19.4	+17.3	+26
Mill towns	-2.3	-7.3	+47
United States	+26.2	+4.7	+53
N.E. region	+20.0	+7.5	+39
All N.E. cities	+5.3	-3.2	+55

*Plus refers to an increase in unemployment rate.

Some observers point to a strengthening of the current manufacturing role. But much of the existing manufacturing base of the 20 industrially mature mill towns in our sample is increasingly becoming cost-ineffective in a highly competitive world economy. These cities must increasingly reckon with the proliferation of Third World manufacturing capacity, as well as with the cost consequences

TABLE 18.9
Changes in Income and Poverty in High-Tech Revitalized Cities and Industrially Mature Mill Towns

	Percentage Changes in Decade of 1970s	
Cities	*Per Capita Income*	*Families Living In Poverty*
High tech	+115	+16
Mill towns	+107	+21
United States	+134	-10
N.E. region	+112	+10
All N.E. cities	+116	+18

of an overvalued dollar. Over time, the result will almost certainly be a continuation of shrinkage in the manufacturing sector—an outcome that increasingly must be anticipated and subsequently addressed.

Other observers see an economic future in nonmanufacturing activity, particularly within the service industries. This, however, seems highly unlikely. New England is a tightly integrated regional economy: The dominant role of Boston is self-evident, and, given the increased importance of the six regional service centers discussed above, it seems highly unlikely that much additional economic activity can evolve from the service sector in mill towns. Nor does such a role seem desirable, given the socioeconomic characteristics that the already-established regional service centers (at least in the southern tier) are assuming.

According to these economic parameters, the outlook for New England mill towns is far from encouraging. We remain convinced, however, that there may be other factors at work that have great power to shape a city's future but are at present only dimly understood. Future researchers should examine, in particular, three additional factors:

- *Commuting patterns.* Countless conversations with high-tech company executives confirm that even though most of their production facilities are not in cities, it is the city-domiciled labor force that they employ in these facilities. Indeed, New England's economic growth for much of the post-World War II period can be characterized by suburban-rural technology investments supported by labor from the regions's declining mill towns. The smallness of New England enables the dynamics of labor markets to work this process out fairly smoothly; but the result contributes little, if anything, to the revitalization of the older cities' economic base, and it most certainly produces a considerable amount of stress on local government's ability to finance itself. Cities that find ways to adjust to or even benefit from these patterns may find a way forward.

- *Physical improvement of the downtown areas*. A number of New England mill towns have substantially renovated their center-city areas, including Providence, New Bedford, Waterbury, and New Haven. Although this approach has upgraded the city's attractiveness, it is only a first step in restoring the fundamental strength of the city's underlying economic base. Physical enhancement in itself is no substitute for the much more important issue of long-run business capital formation. It may, however, be an important indicator of an even more important factor—community attitudes—that is discussed further on.
- *Community attitudes toward change and growth*. What is at stake here is the seemingly ever present conflict between the widely held business perception that an investment in a central city is less attractive (more cost ineffective) than one in a suburban or rural location and the position that positive central city leadership attitudes are strong enough to overcome this negative perception.

To better understand the extent to which positive central city leadership attitudes can overcome negative perceptions of the central city as an investment site, we undertook face-to-face interviews with community leaders and newspaper publishers and editors in eight cities in the region. In the balance of this chapter we will discuss some of the insights based on the role of newspapers in two of these communities, Lowell and Portland.[1]

Clearly the best example of positive community attitudes overcoming economic adversity is in Lowell. Located approximately 20 miles north of Boston, Lowell is the envy of every industrial city in the country. Yet since World War II and until the recent electronics revolution, Lowell suffered from chronic unemployment, abandoned factories, and deteriorating services. Throughout these difficult years, the city never lost pride in itself and *Lowell Sun* editor, Kenneth Wallace, a native of the city, thinks that the people themselves are responsible for the affirmative attitude. He states:

> People's roots are particularly deep in this city. There is a tremendous loyalty to neighborhoods. People find it hard to leave the city. There

> was resentment, however, against the mills. People felt that the mills should be forgotten and perhaps torn down.
>
> But we always had a positive attitude in the editorial page. And when Pat Mogan came along with William Taupier and Joe Tully, we came up with a financial plan to encourage industry and a symbol, the National Heritage Park, to solidify and revitalize community spirit once more.
>
> The creation of the park and the new business that came into the city was supported totally by the *Sun*. This was passed on to our readers because we get into the majority of homes in the Merrimack Valley every day. We talk to people. And you can help a city only if the local newspaper supports the community.

The *Lowell Sun* has always had a positive attitude toward the city. Indeed, some would call it provincial, parochial, and exclusionary. But it seems to have paid off. Lowell is home to one of the major high-technology firms in the world, Wang Laboratories. Also within its corporate borders are two other international firms, Prince Macaroni and Sullivan Printing. The activities of all three firms are well covered by the *Sun*. Wang Laboratories, the major employer in the city and the surrounding communities, gets considerable recognition. Says Wallace, perhaps without qualification but with understandable pride: "Business wouldn't get this type of treatment anywhere else."

Lowell is, therefore, a prime example of aggressive community leadership—helped along by active newspaper support—that has made a notable difference in the city's economy.

Portland is located 103 miles north of Boston. When the Union Railroad station was torn down, it was recognized as the destruction of a charming and attractive symbol that was reminiscent of Portland's past. When the destruction of the old Roman-columned post office followed, Portlanders, aided and abetted by the *Portland Press-Herald*, started an effort to renovate—not destroy—local structures that still had useful life.

"The result," *Press-Herald* editor John K. Murphy says, "is that Portland is enjoying a revival. As a result of the high cost of living in Boston, Portland is in the middle of a renaissance."

It has a new Museum of Art, a new symphony orchestra, a public school system that has won five national awards for excellence and a growing urban real estate market. A positive expression of this was when Great Northern Paper recently moved its headquarters from

Stamford, Connecticut, to Portland. In short, as in the case of Lowell, Portland has benefited from positive community attitudes—in part shaped, and in part reinforced, by an active local newspaper.

CONCLUSION

Our economic analysis reveals that high technology has not been the important driving force in the economies of the vast majority of New England cities. Indeed, only about 1 out of 4 of the 35 cities has been successful in attracting technology-based companies in its revitalization efforts. Although this conclusion does not bode well for the future of older cities, it would be a mistake to write off these cities. Much can be said of the improvements in cities such as New Bedford and New Britain. And if there is any merit in the view that fortune rewards the prepared, then building positive community attitudes—at least, in part through aggressive media leadership—may be as important as anything else in building livable and economically viable cities.

NOTE

1. These interviews were carried out by veteran journalist Paul Pollock, and included newspaper executives as well as local leadership in the following communities: Lowell, Massachusetts; Providence, Rhode Island; New Bedford, Massachusetts; New Britain, Connecticut; Rutland, Vermont; Springfield, Vermont; Portland, Maine; and Claremont, New Hampshire.

REFERENCES

AVAULT, J. (1983) Boston's Development Boom: Economic, Fiscal, Social and Neighborhood Impacts, 1975-83 and 1983-86. (March 1983) Boston: Boston Redevelopment Authority.

BROWNE, L. (1984, April) "High technology and regional economic development." Economic Indicators (Federal Reserve Bank of Boston).

Council for Economic Action, Inc. (1983) The Golden Corridor. New York Times Supplement (February).

HARRISON, B. (1984) "Regional restructuring and good business climates: the economic transformation of New England since World War II," in L. Sawyers and W. K. Tabb (eds.) Sunbelt/Snowbelt: Urban Development and Regional Restructuring. New York: Oxford University Press.

HOWELL, J. M. (1984a) "The role of high technology in the U.S. economy: the New England experience." Speech presented to the Security Analysts Association of Japan, February 20.

——— (1984b) "Formula for growth: the Japanese model versus the New England experience." Economic Review, Bank of Boston.

——— (1984c) "The New England economy and the presidential election: what are the issues?" Seminars on Economic Policy and Presidential Politics, Bank of Boston, October.

——— (1985) "The factors that played a dominant role in the establishment of a high-tech industry in New England and some lessons for other areas." Speech delivered to the Committee for Economic Development of Australia, Melbourne, March.

——— and L. FRANKEL (1983) Historic Mills and Economic Development: The New England Experience. Council for Economic Action.

STEVENS, B. H. (1983) "Shifts in the regional economic outlook and industry growth patterns." Unpublished manuscript.

U. S. Bureau of the Census (1970) Census of Population, General Social and Economic Characteristics.

——— (1980) Census of Population, General Social and Economic Characteristics.

19

Transforming a Manufacturing City: Akron's Redevelopment Plans

FRANK J. COSTA, JACK L. DUSTIN, and JAMES L. SHANAHAN

□AKRON, OHIO, IS OFTEN cited as the quintessential "company" town populated primarily by working-class families who frequent bingo parlors, bowling alleys, and the neighborhood bar when not working the swing shift at the local rubber mill or keeping up with domestic chores.

The facts, however, describe a different kind of place. Over the years, manufacturing employment in Akron declined from 46% of total employment in 1964 to 28% in 1983. Even though rubber tire production—which accounted for much of the manufacturing employment—left the area, the big companies that once produced the tires stayed. As these rubber companies diversified through the years, expanded their facilities nationwide, and grew into multinational corporations, they all made major investments in Akron for conducting central administration, research, and product development. Regarding other firms in the region, many growing ones and even some declining ones are concerned with sophisticated product production. Consequently, Akron's economic future should follow two paths: one, as a place to carry out corporate headquarter functions; and second, as a place for small and medium-size manufacturers that form and prosper from innovation activities.

During the 1970s such certainty about Akron's economic future was not always evident. In fact, the city and its region went through many years of economic self-doubt and a general perception of living through crisis conditions. The intent of this chapter is to describe the transformations and the crises affecting the regional economy and the emerging redevelopment plans that propose to connect an interrelated set of economic activities for the urban core. From this process a new sense of economic purpose has developed that has a potential for renewing the local economy. In some ways Akron literally is building a bridge in this transitional period to a new economic future.

AN ECONOMIC INFRASTRUCTURE THAT ENDURES

In the period from 1870 to 1920 the technologically linked steel, automobile, and rubber industries emerged as powerful forces in the nation's economy that brought significant changes. In spatial terms these industries raised northeast Ohio to a national and world manufacturing center.

By 1920 Akron-based rubber companies successfully captured most of the market for automobile tires and converted the once small city into the rubber capital of the world. Along with steelmaking companies in neighboring cities, the large-scale tire/rubber production plants that clustered in Akron created important economies for their operations and for the region.

But the driving force behind the growth in rubber production came from the automobile industry in Michigan, not more than 150 miles away. The presence of these three national industries created locational advantages for important supply and support companies. Together the agglomeration of these industrial functions developed an economic infrastructure that still rivals any location in the United States.

Quickly following the establishment of large-scale rubber production plants in Akron, new companies formed to support and supply these operations with molds, dies, and various processing and material handling machinery. In turn, over time other companies rose to supply and support these industries, too, such as metal forging and stamping. Many of these new firms were spun off by entrepreneurs formerly connected with the rubber companies, creating a strong industrial network.

Much of this development occurred in the initial decades of growth in rubber production and gave Akron the image of a manufacturing city. However, this birthing process continued throughout the years and, in recent years, has become more specialized and sophisticated.

Today, Akron's economic functions and employment base are both growing and changing. Economic activities once dominated by tire production have given way to corporate functions of administration and research and to education, government, business, and health services.

Akron's workforce has undergone dramatic change in structural form. Manufacturing employment has declined from a high of just over 100,000 jobs in 1953 to roughly 65,000 jobs in 1983. Despite this decrease in manufacturing jobs, total jobs in the Akron SMSA have *increased* during the last three decades, reaching a high of nearly 260,000 in 1979.

Job growth in Akron kept pace with the nation's until 1972. In the four years following the 1974–1975 recession, job formation in the Akron SMSA grew at less than the national rate. The absolute number of jobs peaked in 1979 with 36% more jobs than in 1964 (see Figure 19.1). Manufacturing employment as a percentage of the total began to decline in 1964 from 46% and reached a low of 28% in 1983. Despite these job losses Akron employment has managed to grow because of sharp increases in services, trade, and government sectors. For example, service jobs have more than doubled since 1964 as can be seen in Tables 19.1 and 19.2.

IMPORTANCE OF NONMANUFACTURING INDUSTRY

The importance of nonmanufacturing firms in generating replacement jobs for the net number of jobs lost in manufacturing has been crucial to Akron's economy. The service-oriented industry sector includes transportation services, communications and public utilities, finance, real estate, insurance, miscellaneous other services (including business and health), and government. Over the last 20 years, Akron's nonmanufacturing base has developed into a broader and, in some cases, deeper mix of service/trade industries that more closely mirrors that of the nation.

Akron's fastest growing industries over the last two decades have been in nonmanufacturing and are by order of growth: business services (51.9%); legal services (48.9%); health services (48.1%); and eating and drinking establishments (43.8%). The yearly average employment growth rate for these industry sectors ranges from 4.4% to 5.2% over the last decade. A very distant fifth-place industry is banking/finance (15.4%), where employment grew an average of 1.5% each year (see Table 19.1).

TABLE 19.1
Comparison of Payrolls and Employment in Selected Growing and Declining Industry Groups in Akron SMSA

SIC	*Industry Group*	*Akron SMSA Growth Rate 1974–1983 (%)*	*Percentage of Total Employment in Akron SMSA, 1982*	*Percentage of Total Payroll in Akron SMSA, 1982*	*Job Quality Percentage of Payroll / Percentage of Employees*
I.	Fourteen Growth Groups				
	A. Growth Faster in Akron SMSA than in Nation (1964–1983)				
26	Paper and allied products	+1.1	.745	.882	1.18
27	Printing and publishers	+5.3	1.321	1.435	1.09
54	Retail food stores	+4.3	3.426	2.162	.63
58	Eating and drinking places	+43.8	7.261	2.078	.29
63	Insurance	+2.9	.797	.894	1.12
65	Real estate	+11.6	.646	.436	.67
73	Business services	+51.9	3.141	1.826	.58
81	Legal services	+48.9	.513	.600	1.17

	B. Growth Slower in Akron SMSA than in Nation (1964–1983)				
48	Communications	-5.7	.909	—	—
49	Elec., gas, sanitary	+1.6	1.800	—	—
57	Retail furniture	-21.9	.609	.431	.71
60–62	Banking/finance	+15.4	2.117	1.770	.84
76	Misc. repair services	-19.9	.503	.409	.81
80	Health services	+48.1	7.494	6.921	.92
II.	Twelve Declining Groups				
	A. Declining in Akron SMSA/Growth in Nation (1964–1983)				
28	Chemicals	-12.6	1.519	—	—
30	Rubber and plastics	-45.0	5.664	6.498	1.15
32	Stone, clay, and glass	-39.7	.872	1.092	1.25
35	Machinery (except electrical)	-28.4	5.138	5.548	1.08
42	Trucking/warehousing	-31.8	3.528	4.762	1.35
52	Building materials	-29.6	.899	.700	.78
53	General merchandise stores	-34.2	2.594	1.127	.43
70	Hotels/motels	-26.2	.799	.288	.36
	B. Declining Faster in Akron SMSA than in Nation (1964–1983)				
20	Food and kindred	-16.2	1.071	1.123	1.05
34	Fabricated metals	-27.2	5.434	6.736	1.24
41	Passenger transportation	-45.3	.094	.056	.60
55	Auto dealers	-28.2	2.217	1.521	.69

SOURCES: Ohio Bureau of Employment Services, Labor Market Information Division; and *County Business Patterns*, U.S. Department of Commerce, Bureau of the Census.

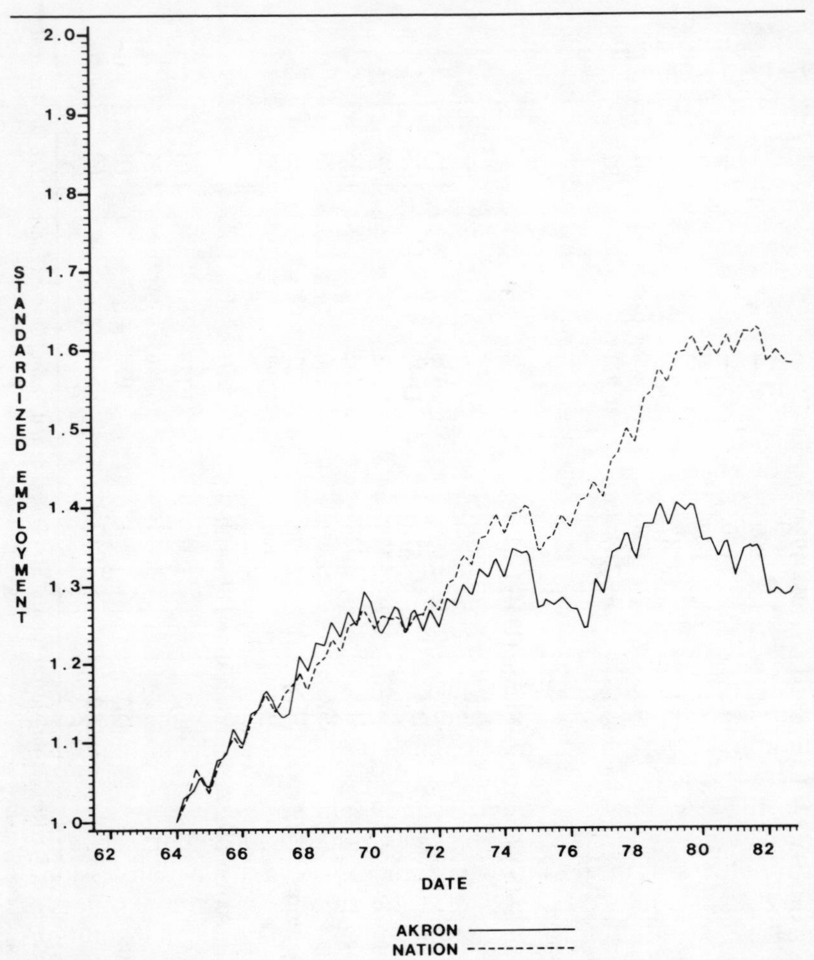

SOURCE: Ohio Bureau of Employment Services and the Bureau of Labor Statistics.

FIGURE 19.1: Total Employment

TABLE 19.2
Industry Change in the Akron SMSA, 1964-1983
(Industry Employment as Percentage of Total Employment)

	1964	*1974*	*1983*	*Change 1964–1983*
Construction	3.3	3.3	2.8	-.5
Mining	.1	.1	.3	.2
Manufacturing	45.9	37.3	27.5	-18.4
Durable	19.8	17.1	13.6	
Nondurable	26.1	20.2	13.5	
Transportation and utilities	6.3	5.7	4.9	-1.4
Wholesale and retail trade	18.8	21.1	23.7	4.9
Finance, insurance, and real estate	3.0	3.2	3.7	.7
Services	10.3	15.1	19.8	9.5
Government	12.3	15.2	17.2	4.9
	100.0	100.0	100.0	0.0

It appears that Akron's place in a service-oriented economy follows from important corporate functions of central administration and research and development. Other strengths of the region include trucking and warehousing wholesale trade in durable goods, electric, gas, and waste services, and retail trade. Given its size and proximity to Cleveland, Akron's performance in support services is strong. Overall, the growth in business services and other professional services as reflected by the increasing number of regional offices of gas and electric utilities, communications, and financial institutions indicates a developing capacity within Akron's service sector to support corporate functions.

IMPORTANCE OF THE MANUFACTURING INDUSTRY

Several sectors of manufacturing show varying degrees of performance in the Akron region. These include printing and publishing and paper and allied products, which are growing faster in the Akron SMSA than in the nation, and fabricated metals, nonelectrical machinery, chemicals, rubber and plastics, electrical machinery, aircraft and parts, and instruments. More specifically, industries where employment growth since 1974 has been good are as follows:

SIC	*New Industries to Akron*
264	Miscellaneous converted paper products
272	Periodicals
275	Commercial printing
278	Blankbooks and bookbinding
281	Industrial inorganic chemicals
284	Soaps, cleaners, and toilet goods
285	Paints and allied products
	Traditional Akron Industries
307	Miscellaneous plastic products
347	Metal services
349	Miscellaneous fabricated metals
354	Metal-working machinery
356	General industry machinery
358	Refrigeration and service machinery

AKRON AS A POLYMER PRODUCTION CENTER

Despite the loss of manufacturing of tires and tubes, significant production of other rubber products remains. Since the 1960s plastics manufacturing has expanded significantly, and together rubber and plastic products have become the region's major source of polymer production.

Most important, Akron continues to be a production center for polymer products and processing technology. Current demand trends call for sophisticated and application-engineered polymer products, which opens up new economic opportunities for experienced people and companies in the region. This advantage is further strengthened by the only major clustering of custom rubber and plastic material processors in the country. These medium to small companies often mix special chemical recipes for producers of specialized rubber and plastic products and for companies pioneering new markets.

Although overall employment in rubber and plastics manufacturing is continuing a downward slide, firms that process *plastic* into products have been growing. By 1982 miscellaneous plastics accounted for 5% of Akron's manufacturing jobs—twice the national percentage—second only to hoses and belts. Today miscellaneous rubber products provide almost one-third fewer jobs than in 1974, a loss that is greater than that occurring nationally. This change can be explained at least partially by the fact that some local firms are mechanizing their operations and converting to plastic products. In short, there are clear signs that low-technology plastics and rubber production in the region is in transition and the shift to new, sophisticated, limited production products caused decreases in local economic employment. However, the long-term effect may be less negative than the numbers suggest.

AKRON'S KEY POLYMER SUPPORT INDUSTRY: MACHINERY

Machinery and mold and tool and die manufacturers started up in the area to support local rubber companies and, to a lesser extent, the steel companies. The history of this set of industries in the region to some extent parallels that of their customers. They grew along with steel and rubber production and expanded their capacities by new technology development, but through the years found it necessary to diversify in order to survive in very competitive markets.

Machinery manufacturing firms have lost 40% of their employment base since 1974 through closure and contraction and contribute only 3% of the manufacturing jobs in Akron regional workplaces. In earlier years companies provided single customers with specialized equipment and services. As time passed, new areas of opportunity opened and technology transferred to new uses, customers, and industries. By the late 1950s, some equipment and technology could be standardized and fell to large centralized production companies. With declining tire production in the region beginning in the 1950s, the diversification of markets and generation of new technology and innovation became critically important. By the late 1970s, two decades of declining regional production, shutdowns, and disinvestment in rubber and steel eliminated those who lagged in new technology and those without sufficient capital to make a transition.

In effect, the region is left with fewer machinery firms; but those that remain for the most part are technologically prepared for the rapid economic changes now occurring. These companies have a business plan built upon developing continuous flows of new technology needed by many different industries.

Mold and tool and die firms have suffered less in the transition and have been a growing industry over the last eight years. They now account for about 5% of Akron's manufacturing jobs. The typical firm has been in business for 17 years, employs 25 workers, and had $1.7 million in sales in 1983. Customers' plants are located in northeast Ohio, elsewhere in Ohio, and in states within a 300 to 400-mile radius. Some sell mostly to customers within this region, but most local firms primarily ship their products out of state.

Both machinery and mold and tool and die manufacturing firms are surviving in Akron, in part because of their nearness to the headquarters of major industrial companies, the reputation and respect that area companies have among their customers locally and in the nation, and the network of businesses that supply these firms with needed materials and technical services.

Overall, manufacturing in Akron, Cleveland, and the surrounding SMSAs in northeast Ohio include both technology-oriented and "low-tech" concerns that are part of a high-caliber technology infrastructure geared to polymer innovations. Developing new products and processing methods are interrelated and are important to understanding the capacity of the region for growth through innovation. The complement of firms extends from chemical suppliers who learned the trade from the bottom up, to custom material mixers who service the small to medium size firms, to machinery and mold and tool and die industries for bringing new products into production, to business services that range from moving products to financing the next new product line. This network of firms has few equals anywhere in the world, yet much of this capacity and knowledge resource waits for someone to make use of it.

CHOOSING A FUTURE IN THE FORCES OF CHANGE

Recently the Office of Technology Assessment concluded that "to an extent state and local initiatives can stimulate the national level of R&D and quicken the pace of commercialization and diffusion of new technologies; they can also contribute to the productivity of the entire U.S. economy" (Office of Technology Assessment, 1984: 4). Arguments and evidence presented in this report and from other authorities contend that broad regions as well as local regions can both benefit from and contribute to technology development and innovation in the United States. Benefits accrue to regions in the form of job growth, lower unemployment, cleaner industry, and an improved tax base. Some regions, because of their economic structure, already function as centers from which technology and innovation initiatives flow and thus have certain developmental advantages.

From the very beginning, it is evident that companies in northeast Ohio were on the cutting edge of innovations in a broad range of polymer-based products for industry and consumer use. In particular, the cluster of major rubber manufacturing companies in Akron led the industry in rubber applications and carried over to plastics and other synthetic products as well. The polymer technology infrastructure, made up of companies, institutions, individuals, and a tradition of entrepreneurship that evolved in this region, developed from a spirit of entrepreneurship and set this region apart from others.

Economic linkages across these organizations were forged by entrepreneurism and caused this region to become a major applied technology center for polymers in the United States. Although today the entrepreneurial spirit is less apparent, this region's technology infrastructure still supports polymer research and development. The comparative advantages for northeast Ohio include:

- a concentration of major corporate research and development facilities primarily engaged in polymer applications, augmented by the presence of private labs and testing services for polymers;
- two universities that are leaders in polymer research and have a record of providing technical services to polymer companies;
- home base for polymer-related industry associations (Adhesives and Sealants Council, Association of Rotational Molders, Polymer Processing Society, and the Polymer Division and Rubber Division of the American Chemical Society);
- access to the corporate decision-making center of the United States for many important manufacturers of polymer products;
- connections to vital polymer material manufacturers in other regions (development, production, and custom mixing of polymer materials);
- a cluster of industrial machinery firms with an industry reputation for solving processing problems encountered in producing complex and innovative polymer-based products;
- clusters of firms with a national reputation for having a rich supply of skilled craftspersons and technicians who can design and build quality molds and dies;
- quality distribution system giving local manufacturers access to a broad region of the United States where the major share of the U.S. market for products made with polymer materials is located;
- and a long tradition of spinning off new polymer applications and companies.

In sum, within the region there exists an advanced technology network for bringing to life new polymer products and production systems. Over the years people within this network gained knowledge and experience essential for technology innovation and precision work. However, in spite of these important resources, the region has lagged in recent decades primarily because it has not yet found new economic activities to replace tire production. In important ways the characteristics of how the local economy developed helps to explain today's painful transition and why the maturing of the rubber industry became a crisis for Akron.

LAND USE AND URBAN DESIGN ASPECTS OF ECONOMIC CHANGE

Economic transformations eventually result in land use change. Land use is the physical expression of a community's economic, cultural, and public interest values. Certain older industrial cities such as Akron exhibit the strength of traditional manufacturing in their land use pattern. Factories and supporting activities have been the dominant elements in the city's land use. Residential neighborhoods such as Goodyear Heights and Firestone Park, oriented to the industrial centers, testify to the strength of the industrial core of the city. In recent decades, however, the old industrial pillars of the city's land use pattern have weakened. The previous section of this chapter describes this process of industrial change. We are here concerned with how the private and public sectors of the Akron community, through conscious efforts in land use planning and urban design, sought to lessen or reverse the adverse economic declines in basic manufacturing.

Land use and urban design planning require public coordination and approval. Large institutions occupying significant amounts of space can and do engage in land use and design planning, but ultimately they must coordinate their efforts with municipal plans for traffic circulation, community facilities location, and land use planning objectives for adjacent areas. The downtown Cascade project was initially marketed as the "International Center for Rubber and Chemicals." City planners sought to accomplish a number of objectives in the redevelopment of the Cascade area: (1) revitalize downtown Akron into the major activity focus of the region, and (2) redevelop the area into an inter-related complex of polymer-related service activities that would help to strengthen Akron's headquarters position in rubber manufacturing. The Goodyear Tech Center's development is another example. During the course of the 1960s and early 1970s, the University of Akron embarked upon a rapid expansion occasioned in large part by the granting of state university status in

1967. Through its powers of eminent domain, the city government assisted the university in its efforts to obtain sufficient land for campus expansion.

All of these expansions and redevelopment activities were carried out with reference to broad general goals of slum clearance, physical and aesthetic improvement, and job creation. In addition, these activities were inserted within the overall plan for community improvement as outlined in such documents as the Land Use Development Guide Plan and the Community Improvement Plan—both guiding policy documents for municipal decision makers. But to a great extent the benefits to be derived from linking these major redevelopment activities were not recognized or understood until much later. Just as the value or benefit of linked organizational responses was not immediately perceived, so too the value and benefit of linked physical development and urban design responses were not immediately seen.

Two important urban design plans covering large areas and linking together adjacent districts resulted in the fairly recent realization that linked physical redevelopment responses should accompany linked organizational responses for economic development.

The first of these is the American City Plan for downtown Akron (see Figure 19.2). This plan, developed by the American City Corporation, seeks to link together a number of important downtown redevelopment initiatives of recent years, including the Cascade Plaza, Quaker Square (an innovative adaptive reuse of a nineteenth-century factory building and concrete silos into shops, restaurants, and a hotel), and the O'Neil's Department Store—the last remaining downtown department store. Planners seized upon the old Ohio Canal as the ribbon that would tie together the various parts of the proposed redevelopment scheme. Perpendicular connectors to the canal in the form of attractive pedestrian walkways and covered skyways would link development throughout the CBD into a network of interconnected spaces. Another important feature of the plan was the goal of linking adjacent areas to the downtown through use of the canal as it extended north and south of the CBD. Thus, a direct connection north through the new Cascade Park toward the Cuyahoga Valley National Recreation Area was envisaged and a connection south to Opportunity Park and beyond was seen as a means to tie these areas together into a lineal development that would increase the attractiveness of the downtown area for many more potential users.

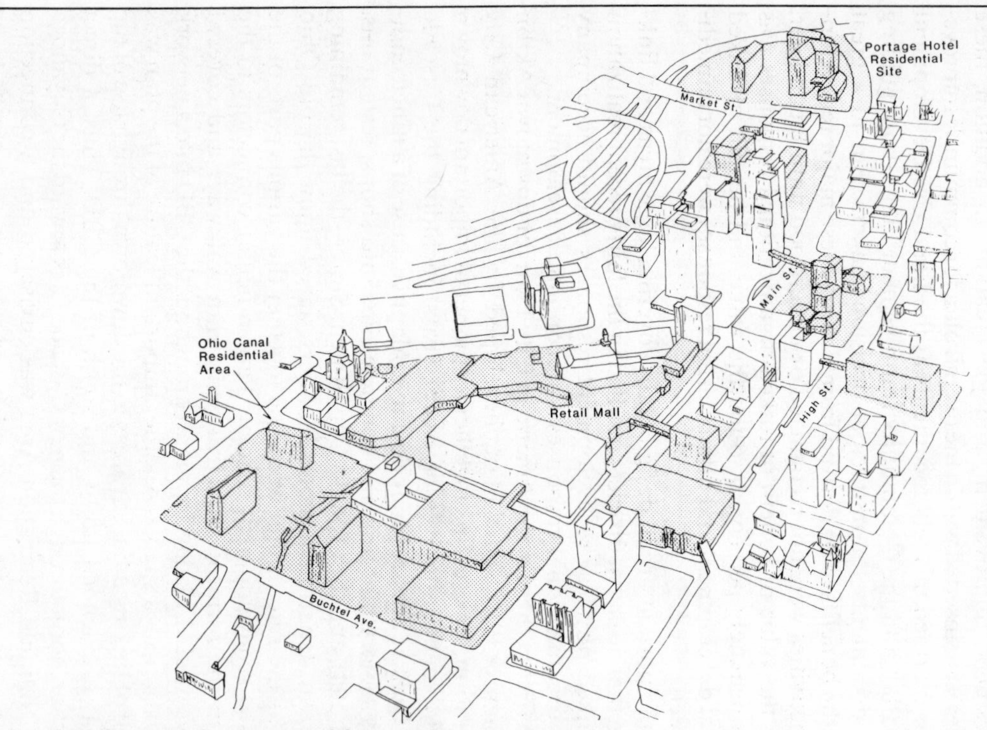

NOTE: New buildings and those proposed for renovation are shown by pattern.

FIGURE 19.2: American City Corp. Plan, 1980

The important points inherent in the American City Plan for downtown Akron include not only a spatially integrated plan for overall redevelopment that seeks to link together important activities within the area, but also the direct links north and south into Akron's neighborhoods, which have the potential of tying these areas together and benefiting the downtown through the attraction of increased numbers of users.

Several criticisms can be made of the plan. One is that it was unveiled as the economy plunged into its deepest crisis since the Great Depression. Its ambitious timetable was simply unrealistic given the economic conditions of the early 1980s.

Another criticism that can be made of the plan is its failure to recognize the economic and physical redevelopment importance of a linkage eastward toward the University of Akron and the Goodyear Tech Center in east Akron. The plan recognizes the importance of the canal as a link to the north and to the south, but the links that can be effected are primarily to recreational and residential areas. These, of course, are very important linkages that have a decidedly favorable impact upon improving overall quality of life in the central city by providing easy access to a variety of recreational, cultural, and retail activities. But, economically important links are those that tie together government, industry, and research activities into product development and product marketing ventures. It remained for a second design study to establish the importance of this link.

"SPAN-THE-TRACKS" LINK

A far more innovative design plan with far-reaching economic development implications is the "Span-the-Tracks" linkage being undertaken by the University of Akron and the city of Akron (see Figure 19.3). The original impetus for this activity came from a planning grant of $250,000 received from the state of Ohio to examine the need for a convocation center in Akron that would meet both University and civic needs for a large, indoor meeting space. The original appropriation for convocation center studies were provided to the state universities located in Akron, Youngstown, and Cleveland. These convocation centers were among the final "bricks and mortar" efforts of Governor James Rhodes—Ohio's "road and public buildings" governor.

From the beginning, Akron viewed the issue of a convocation center differently from other cities. There had long been an expressed

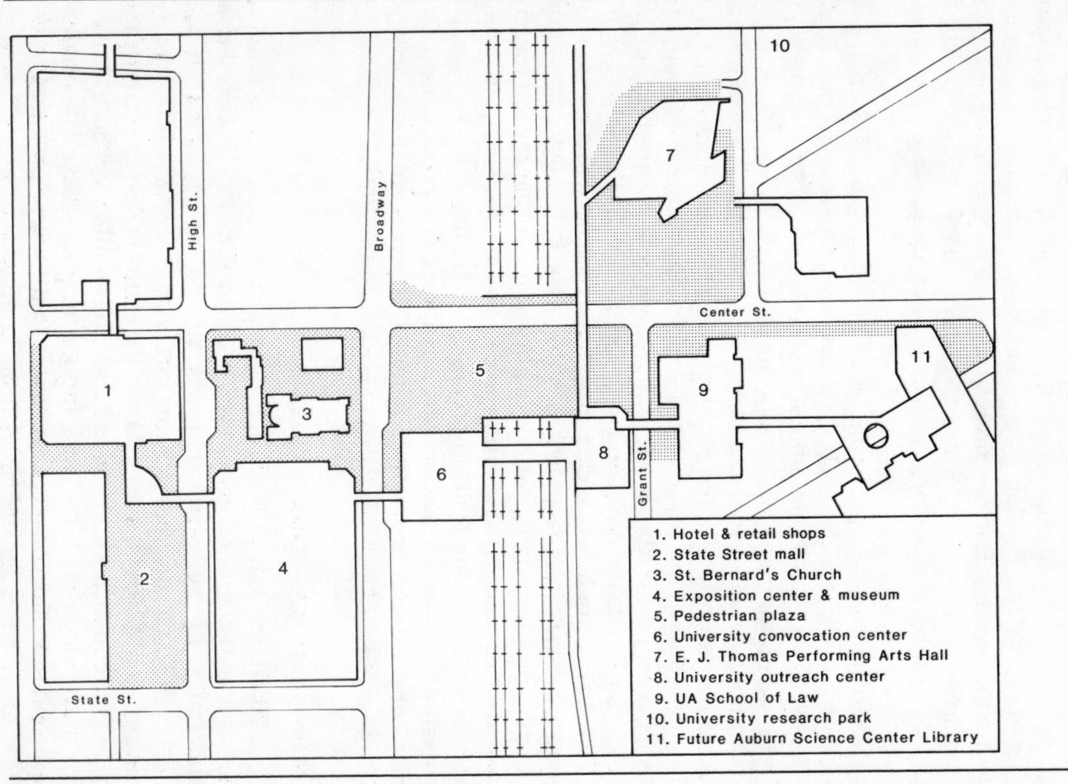

FIGURE 19.3: "Span-the-Tracks" Linkage

goal within both university and municipal planning to "bridge" the physical distance between the campus and the city's downtown by some kind of connector that would create a strong mutual attraction between the two districts. The architect/consultant for the feasibility study, Richard Fleischman of Cleveland, based his design scheme for the convocation study upon the objective of creating a physical linkage between the two districts. His plan utilizes existing structures (some of which are vacant) as interrelated elements along a linkage axis. Linkage ideas in the past had not been based upon functional interrelationships along the path of the link or connector. Attractive ideas of wide pedestrian concourses were devised that were believed to be sufficient in themselves to create a successful link. No specific set of interrelating functions were being combined. Fleishman's plan recognizes the need to envelop these functions in a physical container.

The potentially far-reaching economic development impacts of the link result from the set of interrelated functions contained within it. These include the range of activities encompassing the evolution of an applied polymer research innovation from the University of Akron's research laboratory to the marketplace. Thus, at one end of the proposed link is the University of Akron's proposed polymer research complex, whereas at the other end is the hotel/convention center/exhibition hall complex where new polymer products will be exhibited to potential buyers. Included within the linkage at mid-points are conference and continuing education facilities as well as the university's Performing Arts Hall. The design plan meets the original intent of the appropriation through utilization of existing facilities and augmenting these so that polymer research and development activities are provided with attractive linked sites functioning as both anchors and magnets ensuring the success of the proposed link.

City planning and urban design history give several instructive examples that provide important historic parallels to the Akron effort. The reconstruction of sixteenth-century Rome was based upon links between major centers of religious pilgrimage. L'Enfant's plan for Washington, D.C., links together government centers with avenues and green malls. Nineteenth-century Paris was rebuilt around a network of boulevards acting as axes connecting major monuments and centers of civic activity. Thus, through varying stages of comprehension concerning the historic importance of urban redevelopment utilizing physically linked activity paths, Akron is now at the threshold of significant renewal. An important aspect of this renewal rests upon the successful integration of developing linkages into a network of activities that are not only research and development-based but also linked to spatial settings that enhance overall quality of urban life within Akron.

Akron's physical character is changing. Large areas of aging and physically decrepit manufacturing land are being transformed into campus-like settings for product research and development. The "downtown" is evolving into a multiuse center in which offices, specialized shops, and cultural activities are the dominant users of space. No longer are aesthetics shunned as wasteful. They are now seen as intrinsic elements to success for renewal efforts. Although defining its physical boundaries more precisely, the University of Akron is contemporaneously striving to link up with adjacent districts, especially the "downtown." All of these physical changes and transformations offer a visible expression to the economic transformations that have preceded them and from which they derive their impetus.

REFERENCES

Center for Urban Studies (1985, March) Polymer Technology, Innovation, and Economic Development: Linking the Futures of the Industry and Northeast Ohio. Akron: University of Ohio.

CHANDLER, A. D. (1959, Spring) "The beginning of big business in American history." Business History Reveiw 33: 1–31.

FRY, H. (1984) "In tribute to the chemists who tame rubber." Rubber and Plastics News (August 13).

GIMPEL, J. (1977) The Medieval Machine: The Industrial Revolution of the Middle Ages. New York: Holt, Rinehart, and Winston.

GRISMER, K. H. (n.d.) Akron and Summit County. Akron, Summit County Historical Society.

LeHERON, R. B. and M. D. THOMAS (1975) "Perspectives on technological change and the process of diffusion in the manufacturing sector." Economic Geographer 51 (July): 231–251

MALEKI, E. J. (1979) "Agglomeration and intra-firm linkages in R&D location in the United States." Tijdschrift Voor Econ. en Soc. Geografie 70, 6: 322–331.

——— (1983) "Technology and regional development: a survey." International Regional Science Reveiw, 8, 2: 89–125.

MANSFIELD, E., M. SCHWARTZ, and S. WAGNER (1981) "Imitation costs and patents: an empirical study." Economic Journal 91 (December): 907–918.

National Science Board (1983) University-Industrial Research Relationships. Washington, DC: National Science Foundation.

Office of Technology Assessment (1984, July) Technology, Innovation and Regional Economic Development. Washington, DC: Author.

PRED, A. R. (1975) "Diffusion, organizational spatial structure, and city-system development." Economic Geography 51 (July): 252–268.

University of Akron and Case Western Reserve University (1984, April) A Proposal for an Advanced Technology Application Center in Polymers.

About the Authors

□*BRIAN J.L. BERRY* is Professor of Political Economy and Founders Professor in the School of Social Sciences, University of Texas at Dallas.

□*JOHN P. BLAIR* is Professor and Chair of the Department of Economics at Wright State University. He is author or coauthor of numerous books and articles on urban development including *The Real Estate Market and Feasibility Study*.

□*TERRY F. BUSS* is Director of the Center for Urban Studies at Youngstown State University. He is coauthor (with F. Steven Redburn) of *Shutdown at Youngstown: Public Policy for Mass Unemployment* (1983). He is working on a book about the long-term effects of a plant closing.

□*KENNETH E. COREY* is a Professor of Geography and Planning, Chairman of the Department of Geography, and Director of the Institute for Urban Studies at the University of Maryland, College Park. He is a coauthor of the third edition of *The Planning of Change*. He researches program planning and management methods. His recent research has been on analysis of urbanization policies and processes in Sri Lanka, and on transactional metropolitan policy planning for Seoul, South Korea.

□*FRANK J. COSTA* is Professor of Urban Studies and Geography at the University of Akron and teaches courses in land use planning and control. He was Director of the Center for Urban Studies from 1977 to 1985.

□*TRAN HUU DUNG* is Assistant Professor of Economics at Wright State University. He received his doctorate in economics from the Maxwell School at Syracuse University in 1978.

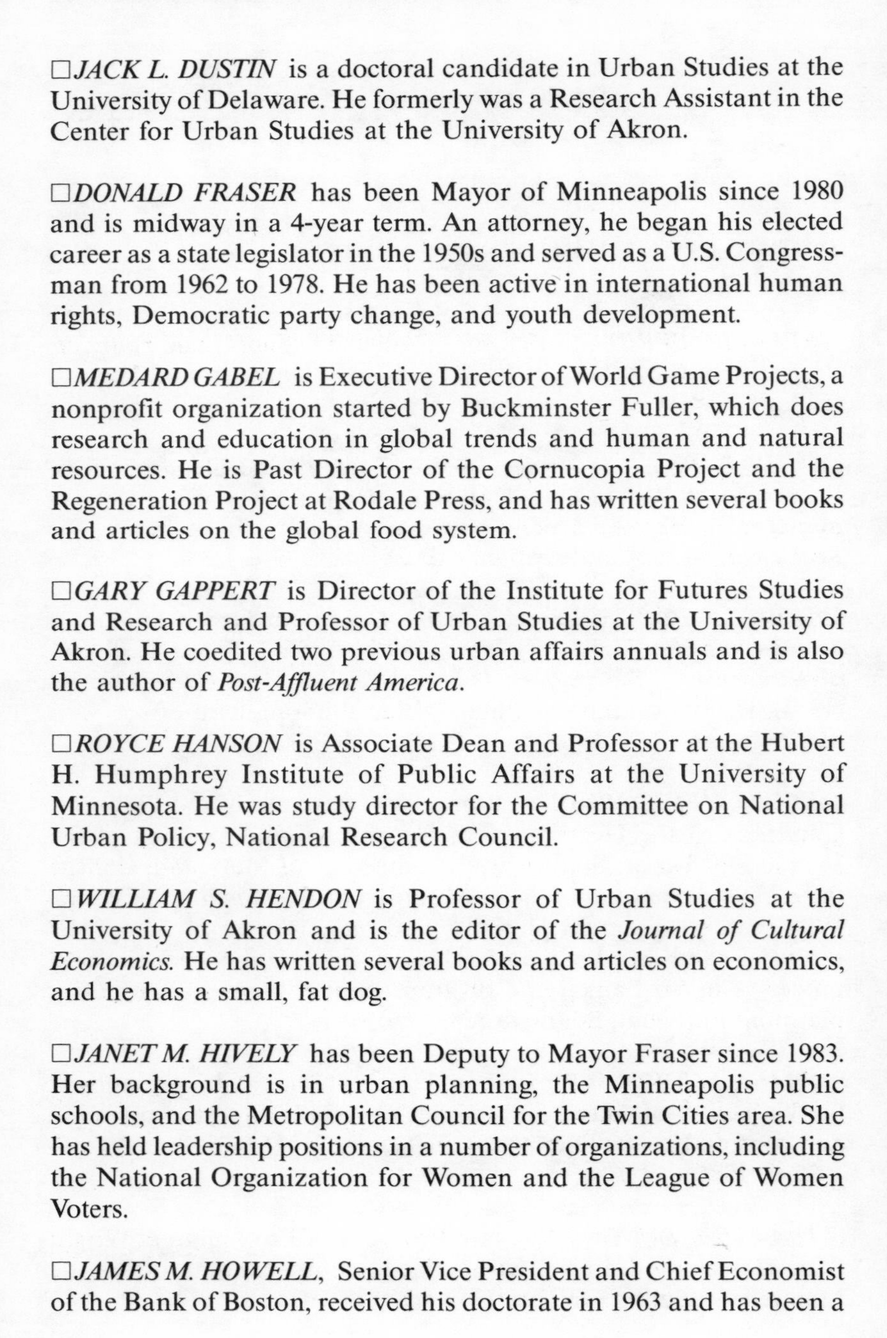

□*JACK L. DUSTIN* is a doctoral candidate in Urban Studies at the University of Delaware. He formerly was a Research Assistant in the Center for Urban Studies at the University of Akron.

□*DONALD FRASER* has been Mayor of Minneapolis since 1980 and is midway in a 4-year term. An attorney, he began his elected career as a state legislator in the 1950s and served as a U.S. Congressman from 1962 to 1978. He has been active in international human rights, Democratic party change, and youth development.

□*MEDARD GABEL* is Executive Director of World Game Projects, a nonprofit organization started by Buckminster Fuller, which does research and education in global trends and human and natural resources. He is Past Director of the Cornucopia Project and the Regeneration Project at Rodale Press, and has written several books and articles on the global food system.

□*GARY GAPPERT* is Director of the Institute for Futures Studies and Research and Professor of Urban Studies at the University of Akron. He coedited two previous urban affairs annuals and is also the author of *Post-Affluent America*.

□*ROYCE HANSON* is Associate Dean and Professor at the Hubert H. Humphrey Institute of Public Affairs at the University of Minnesota. He was study director for the Committee on National Urban Policy, National Research Council.

□*WILLIAM S. HENDON* is Professor of Urban Studies at the University of Akron and is the editor of the *Journal of Cultural Economics.* He has written several books and articles on economics, and he has a small, fat dog.

□*JANET M. HIVELY* has been Deputy to Mayor Fraser since 1983. Her background is in urban planning, the Minneapolis public schools, and the Metropolitan Council for the Twin Cities area. She has held leadership positions in a number of organizations, including the National Organization for Women and the League of Women Voters.

□*JAMES M. HOWELL,* Senior Vice President and Chief Economist of the Bank of Boston, received his doctorate in 1963 and has been a faculty member of George Washington University, the University of Maryland, the Tulane University. He has held the position of Economic Advisor to the assistant secretary of the federal reserve system and has served in an advisory capacity on urban economic development to a number of government agencies. He is currently Chairman of the Council for Economic Action, which conducts research and undertakes projects involving urban revitalization.

□ *KNOWLTON W. JOHNSON* is Director of the Urban Studies Center in the College of Urban and Public Affairs at the University of Louis ville, Louisville, KY. He received his doctorate in social science from Michigan State University in 1971. His publications and technical reports include research in the areas of social indicators and planned change.

□ *RICHARD V. KNIGHT* is an Economist specializing in the development of city regions. He is a professor at Antioch College, Yellow Springs, OH.

□ *C. THEODORE KOEBEL* is Associate Director of Research at the Urban Studies Center, College of Urban and Public Affairs, University of Louisville, KY. His is also Director of Housing and Economic Development Research. His doctorate in urban planning and development is from Rutgers University. He has conducted other research focusing on urban demographics and economic development issues.

□ *JORMA MÄNTY* is Associate Professor of Urban Planning at Tampere University of Technology, Finland. He has been working as Urban Planner in the city planning offices of Helsinki and Vantaa and as Division Chief in Helsinki.

□ *YASUO MASAI* is Professor of Geography at Rissho University in Tokyo and is Chairman of the International Cartographic Association Commission on Urban Cartography. He is the author of *Geography of America and Canada—A Japanese View, A Comparative Study of Japanese and American Cities,* and *Urban Environment and "Japan."*

□ *HAROLD M. MAYER* is Professor of Geography at the University of Wisconsin-Milwaukee, and Vice President of the Milwaukee Board of Harbor Commissioners.

□*GEORGE PALUMBO* is Professor of Economics at Canisius College and spent 1984–1985 on sabbatical leave at Syracuse University. His fields of specialization are urban and regional economics and state and local finance.

□*NORMAN PRESSMAN* is Associate Professor of Urban and Regional Planning, University of Waterloo, Ontario, Canada. He is cofounder of the Livable Winter City Association in Canada. He has written numerous articles and papers in a broad spectrum of international journals.

□*MICHAEL L. PRICE* received his doctorate in sociology from Michigan State University. Since 1981 he has directed the Population Studies Program at the University of Louisville, Louisville, KY. He has conducted other research in the areas of demographics and migration's impact on populations.

□*F. STEVENS REDBURN* is Senior Policy Analyst at the U.S. Department of Housing and Urban Development. He is coauthor (with Terry F. Buss) of *Mass Unemployment: Plant Closings and Community Mental Health*. He is now working on a book about discouraged workers.

□*ROBERT ROSS* is an Economist with the Farmers Home Administration of the U.S. Department of Agriculture.

□*SEYMOUR SACKS* is Professor of Economics in the Maxwell School of Syracuse University. He has served as a consultant for government and business in school finance, municipal finance, and economic database analysis.

□*SUSAN W. SANDERSON* is Senior Research Associate in the Human Resources Program Group at The Conference Board. She was formerly Adjunct Assistant Professor of Social and Economic Development at the School of Urban and Public Affairs, Carnegie-Mellon University.

□*JAMES L. SHANAHAN* is Professor of Urban Studies and Director of the Center for Urban Studies at the University of Akron.

□*DOUGLAS V. SHAW* is Associate Professor of Urban Studies at the University of Akron and is the author of *The Making of An Immigrant City: Ethnic and Cultural Conflict in Jersey City, New Jersey, 1850–1877.*

□*SHELBY STEWMAN* is Associate Professor of Demography and Sociology at the School of Urban and Public Affairs, Carnegie-Mellon University.

□*JOEL TARR* is Professor of History and Public Policy at the School of Urban and Public Affairs, Carnegie-Mellon University. His current research focus is on long-term environmental problems, urban infrastructure, and regional economic change.

□*PHILIP R. THOMPSON* is a Research Associate in the College of Urban Affairs at Cleveland State University. He earned his doctorate in anthropology from Rutgers University in 1979 and his Master's of library science from the University of Michigan in 1982. He is the author of 5 scholarly articles, 3 popular articles, and 8 other items. He is the coauthor of 2 other articles on urban economic development.

□ *WILBUR R. THOMPSON* is a principal of Thompson Associates, a consulting firm specializing in state and local economic development strategy, located in Albuquerque, NM. He is the author of *A Preface to Urban Economics* and other works on the revival of industrial cities.

□*JON VAN TIL* is Associate Professor of Urban Studies and Public Policy at Rutgers University—Camden College and is author of *Living With Energy Shortfall: A Future for American Cities and Towns* (1982). During the 1986-1987 year, he is Visiting Professor of Sociology at the University of Colorado, Boulder.

□*XENIA ZEPIC* is a Research Fellow at the Institute of Urban Studies, University of Winnipeg, on leave from the Metropolitan Toronto Planning Department, where she has been working as a planner/urban designer. She is cofounder and President of the Livable Winter City Association.

Other available
URBAN AFFAIRS ANNUAL REVIEWS . . .

A semiannual series of reference volumes discussing programs, policies, and current developments in all areas of concern to urban specialists.

THE DELIVERY OF URBAN SERVICES: Outcomes of Change (Vol. 10)
Edited by Elinor Ostrom

URBANIZATION AND COUNTERURBANIZATION (Vol. 11, in cloth only)
Edited by Brian J.L. Berry

MANAGING HUMAN RESOURCES: A Challenge to Urban Governments (Vol. 13)
Edited by Charles H. Levine

THE RISE OF THE SUNBELT CITIES (Vol. 14)
Edited by David C. Perry and Alfred J. Watkins

ACCOUNTABILITY IN URBAN SOCIETY: Public Agencies Under Fire (Vol. 15)
Edited by Scott Greer, Ronald D. Hedlund, and James L. Gibson

THE CHANGING STRUCTURE OF THE CITY: What Happened to the Urban Crisis (Vol. 16)
Edited by Gary A. Tobin

FISCAL RETRENCHMENT AND URBAN POLICY (Vol. 17)
Edited by John P. Blair and David Nachmias

URBAN REVITALIZATION (Vol. 18, in cloth only)
Edited by Donald B. Rosenthal

RESIDENTIAL MOBILITY AND PUBLIC POLICY (Vol. 19)
Edited by W.A.V. Clark and Eric G. Moore

URBAN GOVERNMENT FINANCE: Emerging Trends (Vol. 20)
Edited by Roy Bahl

URBAN POLICY ANALYSIS: Directions for Future Research (Vol. 21)
Edited by Terry Nichols Clark

URBAN POLICY UNDER CAPITALISM (Vol. 22)
Edited by Norman I. Fainstein and Susan S. Fainstein

CITIES IN THE 21st CENTURY (Vol. 23)
Edited by Gary Gappert and Richard V. Knight

THE GREAT HOUSING EXPERIMENT (Vol. 24)
Edited by Joseph Friedman and Daniel H. Weinberg

CITIES AND SICKNESS: Health Care in Urban America (Vol. 25)
Edited by Ann Lennarson Greer and Scott Greer

CITIES IN TRANSFORMATION: Class, Capital, and the State (Vol. 26)
Edited by Michael Peter Smith

URBAN ECONOMIC DEVELOPMENT (Vol. 27)
Edited by Richard D. Bingham and John P. Blair

HIGH TECHNOLOGY, SPACE, AND SOCIETY (Vol. 28)
Edited by Manuel Castells

URBAN ETHNICITY IN THE UNITED STATES: New Immigrants and Old Minorities (Vol. 29)
Edited by Lionel Maldonado and Joan Moore

CITIES IN STRESS: A New Look at the Urban Crisis (Vol. 30)
Edited by M. Gottdiener

NOTES

NOTES

NOTES